ERWIN STRAUS *(Cir., 1930)*

DUQUESNE STUDIES—PSYCHOLOGICAL SERIES

GENERAL EDITOR—AMEDEO GIORGI

Department of Psychology, Duquesne University

VOLUME TEN

Man,
Time,
and World

Two Contributions to Anthropological Psychology

by ERWIN STRAUS
translated by Donald Moss

DUQUESNE UNIVERSITY PRESS

PITTSBURGH

Published by:
Duquesne University Press
600 Forbes Avenue
Pittsburgh, PA 15282

Distributed by:
Humanities Press
Atlantic Highlands New Jersey 07716

First Edition

Library of Congress Cataloging in Publication Data

Straus, Erwin Walter Maximilian, 1891–1975.
  Man, time, and world.

  Translation of: Geschehnis und Erlebnis and Der Archimedische Punkt.
  Bibliography: p.
  1. Psychology—Philosophy.   2. Psychological research.
3. Time—Psychological aspects.   4. Traumatic neuroses.   I. Straus,
Erwin Walter Maximilian, 1891–1975.   Archimedische Punkt.
English.   II. Title.
BF38.S6813   1982   616.89   82–12946
ISBN 0–8207–0159–9

Contents

v

Translator's Preface

I. ERWIN STRAUS, THE MAN AND HIS LIFE

Erwin Walter Maximilian Straus was born in Frankfurt am Main on October 11, 1891, the second son of Caesar Straus, a banker and social philanthropist, who had introduced the idea of housing cooperatives for laboring men into Germany. His mother, Antonie Straus-Negbaur, was a Brooklyn born woman of German parents, and a renowned art collector.

Erwin Straus received a comprehensive classical education—including attendance at the Lessing Gymnasium in Frankfurt—at a time when German learning was at a high point. This education left him imbued with the belief that the one true revolution in human thought was that of Greece in the classical period. He was also a part of that especially gifted and yet alienated generation of German-Jewish intellectuals, which Hannah Arendt has so penetratingly described in relation to Straus' contemporary, the critic Walter Benjamin. It was their curse and their gift simultaneously to possess a keen awareness of the significance of history, tradition, and time itself, and to observe the dissolution of whatever historical order they themselves sought to inhabit (Arendt, 1969).

In 1910 Straus turned to the study of medicine—and philosophy. In the following years he attended the lectures of Carl Jung in Zurich, of Husserl and Reinach at Göttingen, of Stumpf and Riedel in Berlin, and of Scheler, Pfänder and Moritz Geiger in Munich. The ethical emphasis in Straus' thought is a reminder of his early affinity for Scheler's viewpoint. Straus' studies were interrupted by military service as a *Feldarzt*, a field doctor, during World War I, including action on the Polish front.

Straus' medical training brought him under the influence of Karl
Bonhoeffer, the great neurologist and father of the "death of God"
theologian, Dietrich Bonhoeffer, of Ludwig Eddinger, another
pioneer in German neurology, and of a third neurologist, Richard
Cassirer, who was both a cousin and a teacher of Straus. (Straus was
also a cousin of Ernst Cassirer, the German philosopher whose works
include his *Philosophy of Symbolic Forms*, and of Bruno and Paul
Cassirer, the noted art handlers and publishers). In the 1920's Straus
formed enduring friendships with the leading figures of the new
movement of "anthropological" psychiatry—Victor von Gebsattel,
Ludwig Binswanger, and Jurg Zutt. With them he founded *Der Nervenarzt*, the journal which served as the voice of existential and
anthropological thought in psychiatry. In this same period, Straus also
formed an acquaintance with the French phenomenological psychiatrist Eugene Minkowski. In 1920 he married Trudy Lukaschik, a
concert violinist and poetess.

These were the years of the Weimar Republic, a period of impressive fecundity in the German arts and sciences, including psychiatry
and psychology; Straus' works bear many traces of this ferment.
Gestalt psychology was strongly represented in Berlin in the 1920's and
1930's, and Straus put a central emphasis on the articulation of part-
whole relations in his early works. Yet, Straus' use of the notion of
Gestalt hearkens back more to Goethe than to Wertheimer. The Greek
Eidos, the Platonic essence, also seems to lurk behind Straus' use of
Gestalt. Throughout his life Straus read and re-read the classics of
antiquity, especially Aristotle, as well as Augustine, Goethe, and
Shakespeare. His psychology has more affinity with the worldview of
Hamlet, of Faust, and above all with that of the Greeks, than it does
with any modern psychology or psychologist.

From the beginning Straus' psychology differed markedly from that
of his fellow anthropological psychiatrists in its point of departure.
Binswanger chose the thought of Husserl, and later of Heidegger as his
starting point, and Minkowski chose Bergson. In contrast Straus,
however rich the sources on which he drew, set forth from direct
reflections on the events of everyday life, as well as on the clinical
experiences of psychiatry. In this method he was truest to the motto of
Husserl, the founder of phenomenological philosophy: "back to the
things themselves."

Herbert Spiegelberg has remarked that Straus was no methodologist

(Spiegelberg, 1972, p.267). In a broader sense, however, Straus exhibited a strict methodology consistent from his earliest major monograph (1925) to his latest essays (1967, 1975). His was a methodology of thinking, of critical reflection, of posing the proper question, and of exposing the ultimate import of a questionable theory. This is the time honored methodology of dialectics, identical with that of Socrates in the *Thaetatus*: ". . . it has been possible for me . . . to develop the course of their ideas further and so to develop for the first time the immanent content of their teachings in its entirety . . ." (Straus, 1925).

*   *   *

Straus' academic career began in a promising fashion; he became Privatdocent for Psychiatry at the University of Berlin in 1927, and Unofficial Extraordinary Professor in 1931. Within two years of Hitler's seizure of power, however, he was forced to cease teaching, and leave the editorial staff of *Der Nervenarzt*. He continued to practice medicine on a limited basis, and participated in a philosophical study group with Jurg Zutt, von Gebsattel, Jacob Klein, and others. Klein, a philosopher who later emigrated to Annapolis, Maryland, recalls Straus' participation in the "Platonic dialogues" of this group (Klein, 1977). This time of ostracism was perhaps the most crucial period in the development of Straus' mature anthropology, with its equal emphases on human temporality and human spatiality. His most renowned full length work, *Vom Sinn der Sinne* (English edition, *The Primary World of the Senses*) appeared in 1935.

Warned of danger in 1938, Straus payed 36,000 Reichsmarks, a "Reichsflucht Steuer," for the privilege of fleeing his homeland. For the next eight years, he and his wife resided in hardship at Black Mt. College in North Carolina. There Straus taught "humanistic psychology," and shared the artistic and educational enthusiasms of the Black Mt. community. Eventually, however, he became disenchanted, believing that such artistic and social experimentalism was the work of those who "welcome the decline or breakdown of the old order as the breakdown of order as such" (Straus, 1941). The schisms and conflicts within the Black Mt. community were vicious at this time, and Straus' conservatism won him the reputation of being a reactionary and even a racist (Cf. the unfavorable discussion by Duberman, 1972). As a more sympathetic colleague put it: Straus was a true Greek, and that implies an aristocratic, elitist attitude. He by no means assumed that all men are equally endowed by nature at birth (Fischer, 1977). He was also convinced that human individuality, in the form so valued in the West, presupposes collective cultural forms and norms, which were seriously endangered in this time of crisis (Straus, 1941).

During the years 1944 through 1946, Straus intermittently conducted research at Johns Hopkins in Baltimore, which later formed the basis for his monograph *On Obsession* (1948). There too he obtained his American medical and psychiatric diplomas.

In 1946, Straus left Black Mt. for Lexington, Kentucky, to become director of Research and Education at the Veterans Administration Hospital there. Lexington became and remained his American home, the center from which he ventured forth on his extensive travels and speaking engagements, in Europe and America. He conducted an ongoing seminar on phenomenological psychology at the VA Hospital, as well as five "Lexington Conferences on Phenomenology, Pure and Applied." From Lexington he played a formative role in introducing this country to the anthropological and phenomenological orientations in psychiatry and psychology.

His writing continued, and included new and original contributions, such as his beautiful monographs on the upright posture (1948 b, 1963 b, 1963 c), as well as returns to the long familiar themes of his youth. His 1975 essay, for example, "The Monads have Windows," returned to a theme of Leibniz which he first picked up in the 1920's: We humans are separate and yet live together in a common world. The image of the isolated subject is refuted not only—as he wrote in his 1925 monograph on suggestion—by the primary pathic bonds of attraction and repulsion linking all human beings; rather there is also a gnostic bond, in the common spectacle of the shared world. To quote the 1926 version: "The monads are indeed all substantially and essentially separate, but they resemble one another nevertheless, insofar as each mirrors the entire universe, although each from a different vantage point and in different degrees of clarity or confusion" (Straus, 1926, p.121). The miracle of human seeing, or better of human beholding, was both Erwin Straus' peculiar personal gift and a universal human endowment. Following his death on May 20, 1975, his gravestone was inscribed with the words of Goethe, words Straus had used as the title of one of his most beautiful essays:

> Zum sehen geboren, zum schauen bestellt.
> Born to see, bound to behold.

II. THE WORKS IN THE PRESENT VOLUME

Context: The two works in the present volume span 27 years (1930-1957) and exhibit, in their earliest and in their mature form, many of the major themes of Erwin Straus' writings. In the remainder of this preface I will sketch out the general context and intent of Straus' early works, and situate these works against that background. My commentary on the translation itself, especially terminological problems, can be found in the accompanying footnotes.

Straus' first *Habilitationsschrift*, *The Problem of Individuality* (1926), provides a programmatic key, through its declarations of principles and intent, to much of his life's work. There he declared the need for philosophy to undertake, "to clearly establish the foundations of the human sciences (*Geisteswissenschaften*) which are independent of the natural sciences" (1926, p.27). There he reflected upon the principles of Goethe's theory of colors, which stand in sharp contrast to Newton's physicalistic science of optics. He cited Goethe in emphasizing how much theorizing is disguised within the supposedly objective observations of psychology: "Because the mere viewing of a thing cannot direct us. Every act of viewing turns into a contemplating, every contemplating into a reflecting, and every reflecting into a connecting, and thus one can say, that already with every attentive glance into the world we theorize" (Goethe, cited by Straus, 1926, p.26).

There too Straus introduced the holistic biological viewpoint which was to distinguish his thinking from that of many of his existential colleagues. He developed a mode of viewing the organism, emphasizing its characteristic temporality—through which an organism preserves its past in its present (1926, p.70). Straus defended the concept of life against reductionism, emphasizing the impossibility of a complete translation of vital processes into mathematical equations or the laws of physics. And, Straus added, "Even more than biology in general, medicine—as a special biological discipline—is bound to the world of appearances in its qualitative fullness" (1926, p.30). It is significant that Straus' 1926 monograph was published in a massive compendium—*The Biology of the Person: a Handbook of the General and Specific Theories of Constitution*—which specifically aimed at defending the holistic and constitutional viewpoints within medicine against the otherwise prevailing forces of mechanism and reductionism. Straus' works were polemical from the beginning, and explicitly aligned him with the widespread movement affecting the sciences and humanities of the time, a movement variously identified with the concepts of Gestalt, of totality, of the person, and of life (1926, pp.27–28).

Finally, in his *Habilitationsschrift*, Straus introduced the historiological viewpoint, which was to color so many of his works. What distinguishes the historically oriented human sciences from the natural sciences, he believed, was not merely the difference in their objects of study, "but rather the distinction between the generalizing natural scientific and the individualizing historical modes of conceptualizing" (1926, p. 64). Like Minkowski, Straus believed that the central question in the various schools of psychology was that of time, of the historical development of the human being:

> In this context everything depends upon how time is conceived. We need a theory of time, which makes it possible to transform the mere succession of a temporal series into a relationship immanent to the things, and which therefore makes it possible to overcome historicism in every form. We must once again bring into play the . . . thoughts of Augustine regarding time, which stand so close to modern theories—e.g., that of Natorp, which holds that consciousness is not within time, but rather time within consciousness; or that of Max Scheler concerning the interpenetration of the manifold—we must once again call upon these thoughts, if we wish to provide the general theoretical foundations for the conceptions of development which are current today in biology and psychology. These concepts have taught us to see the facts anew and shown us the way to the discovery of new facts (cf. for example, the transformation in the understanding of learning, which has taken place in child psychology, e.g., with Koffka). Only through the interpenetration of the manifold will the genetic understanding become an understanding in the genuine sense (Straus, 1926, p.118).

Let us turn, against this general background, to consider the works in the present volume.

Event and Experience (1930):

In this bold and far reaching work Straus sets out from the facts of the compensation neurosis, a disorder involving the claim that an individual has been disabled through a "psychic trauma," and is therefore entitled to a compensatory pension or settlement. This narrow clinical base provides a vehicle through which Straus calls into question and

does battle with the then most influential schools in psychology—
Freudian psychoanalysis and Pavlovian reflexology. Further, Straus
endeavors to put forth a comprehensive theory of his own—the historiological perspective—broad enough to account for and more
adequately comprehend the findings of both genetic schools. This
project was bound to fall short in some sense, and drew criticism even
from his close colleague Binswanger (1931). Nevertheless, Straus
succeeds in disclosing some of the barest conceptual threads in the
fabric of the genetic psychologies, and sketches out the outline of a
phenomenologically adequate general psychology, based upon a close
attention to the structure of human temporality.

We encounter here an abundance of Straus' methodological and
critical observations: For example, he attacked the "psychology of
dearth," i.e., the propensity of psychologists to use examples and
experiences robbed of their fullness and relational richness. Just as the
student of art must study masterpieces, the student of psychology must
study human experiencing in its completeness and richness. So too,
when drawing examples from pathology, the researcher must incessantly inquire: ". . . how must the undisturbed behavior be constituted, if pathological phenomena of this kind are to be possible?"
(Straus, 1930). As Straus had earlier observed, "Only the man who
carries in himself a virtual image of the intact whole is able to perceive a
torso" (1926, p.123).

Walter Bräutigam, in a preface included in the present English
edition, has shown the unmet challenge which *Event and Experience*
presents to the theoretical debate within psychoanalytic circles.
Elsewhere, I have demonstrated the cutting edge of the critical principles enunciated here, when applied to present day neuropsychology,
and especially to the theories of Karl Pribram (Moss, 1981 a).

Each of the critical polemics in *Event and Experience* is subordinated to the encompassing goal of providing adequate theoretical
foundations for the conceptions of development implicit in the genetic
psychological theories. Thus it is here, in *Event and Experience*, that
Straus fulfills the promissory note of his *Habilitationsschrift*, and
pursues a systematic anthropological account of the human being as *ein
Werdender*, an individual in the process of becoming. To cite one
critical passage: "The consciousness of the individual person unfolds
as the experience of his own inner history. Every single moment is a

phase of his inner history. Everything coming into consciousness in a specific moment, is determined by how it fits into this becoming, or how it arrests or runs contrary to it" (1930).

Definite existential themes permeate the monograph. For Straus, the human experience of time *is* the basis for the moral experience. Underlying the detailed consideration of the psychic trauma in this work, is the ethical question: how does the fortuitous *Geschehnis*, the outward fact, fatefully compel an *Erlebnis*? What is the individual's deeper involvement and responsibility, in the appropriation of his unsought destiny? How are we to reconcile individual freedom with an experienced compulsion?

Straus discloses an ethical element within all human experiences— both normal and morbid—by viewing adventurous behaviors, romantic suicides, phobic symptoms, and even the perversions as expressions of one's relation to the totality or absolute: ''. . . every experience and every structuring of life, viewed as a whole, takes up a position toward, and answers, the existentiell question which continuously frames every experiencing'' (Straus, 1930). It is in this sense that he introduces the notion of the *existential neurosis*, as the failure to meet the challenge of a new phase in life, the exposed existence of the adult.

Event and Experience is marred by several minor conceptual flaws, which I will briefly mention here. I refer the reader for greater detail to the discussions by Binswanger (1931), Boss (1947), and Moss (1978, 1981 b). Binswanger, and later Boss deplored the concept of *deformation*, by which Straus characterizes the perversions. Both Binswanger and Boss preferred to place in the foreground, as the basis for the perversions, a disturbance in the experience of community, a disturbance of our being-with-one-another. Both saw the perversions, and pathological behavior in general, as desperate efforts by an isolated individual to actualize himself.

Binswanger also contested the accuracy of Straus' appropriation of Binswanger's own notion of the ''inner life history,'' and objected to the needless proliferation of concepts, such as historical *modality*, and representative *function*—the latter a term Straus borrowed from his cousin Ernst Cassirer. Further, Binswanger felt that the key concepts of self-actualization and self-abandonment involved an unclarified concept of self, which did not cohere well with Straus' general emphasis on the historically unfolding forms of actuality (Dilthey). He delighted, on the other hand, in several other terms which represented purely phenomenal "movement concepts"—"Letting oneself fall into decline," "to linger," and "to sink into absorption in . . . " It is worthy of note that in Straus' later works the language of movement, posture, and position predominates.

The Archimedean Point (1957):

There is considerable continuity between *Event and Experience* and this later work. Both the 1930 monograph and the 1957 essay explore the relationship between the world and the inner life history, between event and experience. In each case Straus refuses to follow either of the typical reductions: 1) that of naturalistic psychology which reduced human time to the Newtonian time of objective world processes, or 2) that of philosophical idealism—and many phenomenologists—who reduce world events to inner human time. "In sensory experience, in the I-world relationship, two temporal orders encounter one another and are differentiated, that of my own becoming and that of the world. I experience my Now as a moment in the occurrence of the world" (1957). I experience the encompassing world as Other—as the Allon.

Binswanger had criticized the seeming dualism of *Event and Experience*, and emphasized the dialectical interrelation between event and experience, and individuality and world (Binswanger, 1931). Straus, however, clung tenaciously to this dichotomous opposition, as expressing a primordial given in our experience of the world. In *The Archimedean Point*, he again emphasizes the indissolubility of this tension: "the relationship I-world is maintained throughout every change" (1957). Further, he declares that the real "Archimedean point" is not outside the world of man, as Cartesian science would have us believe, but rather, within the polarity of the I-world relationship. Man, in his motility, orients himself beyond the actual toward the possible, and experiences the world as the encompassing, the mighty, and the enduring. In this encounter with the world, the human being possesses the basis for becoming *homo faber*, the builder.

In *The Archimedean Point*, Straus goes beyond the earlier viewpoint of *Event and Experience*, through a presentation of his mature anthropology, which links the structure of man's sensory experience to his motility and upright posture. As the temporal horizon ordering every experience was the theme of *Event and Experience*, here it is the spatial horizon, shaped by our embodiment. Straus describes first the structure of our action-space, which is organized around our pathic involvement with a situation of scarcity and want. He also, however —preparing the way for his later essay, "Born to See, Bound to Behold" (1963 b)—introduces the nature of our gnostic involvement in an aesthetic realm. In man's encounter with the firmament, Straus finds a paradigm for the astonished wonderment of the aesthetic attitude. The starry sky invites man to relate himself, in contemplative detachment, to an encompassing whole.

In the process of Straus' exposition, familiar critical themes reappear. Neuropsychological theories, he points out, do violence to behavior by stripping it of its human "subjective" attributes, and in turn compensate for this by attributing anthropomorphic qualities to brain function. This essay reminds us that the goal of Erwin Straus' scathing and unceasing polemics against naturalistic psychological theories was never simply greater objectivity and truth in a knowledge-theoretical sense. Rather, the anthropological strain is present throughout his works (Moss, 1981 b). Naturalistic theories are an affront to the dignity of man, the actor and individual (Straus, 1941). The stature of man the scientist is diminished by the propagation of the seemingly scientific S-R theory, which would reduce the entire human drama to the level of the sunburn reaction of skin when exposed to light (Straus, 1963). Straus' critical essays thus take their place alongside his anthropological studies, such as "The Upright Posture" (1948 b), "Man: A Questioning Being" (1953), "The Sigh" (1952), and "Born to See, Bound to Behold" (1963 b).

We may close with one of Straus' most explicit statements on the paradoxical nature of the human animal:

> As individualities we belong to nature, as individualities we belong to a spiritual objective order. As individuals we are marked by some peculiarity, such as the fingerprint; we become individualities insofar as we integrate objective orders and adapt ourselves to them. As individualities we are specimens of a zoological species, and we are restrained to the present in space and time. As individualities we are in a potential relation toward the whole of the world, to the past and to the future (Straus, 1941).

DONALD MOSS

TRANSLATOR'S PREFACE

REFERENCES

In the preceding text all works have been cited by original publication dates, in order to preserve a sense of chronology. In the following references the version of the text actually cited will be given a full reference; the original publication date, if different, will appear last in parentheses.

Arendt, H. "Walter Benjamin:1892–1940." In Benjamin, W. *Illuminations*. N.Y.: Schocken, 1969.
Binswanger, L. "Geschehnis und erlebnis, zur gleichnamigen schrift von Erwin Straus." *Monatschrift für Psychiatrie und Neurologie*, 1931, 80, 243–273.
Boss, M. *The meaning and content of sexual perversions*. N.Y.: Grune and Stratton, 1949. (1947)
Duberman, M. *Black Mountain: an exploration in community*. N.Y.: Dutton, 1972.
Fischer, W. Personal communication, 1977.
Klein, J. Personal communication, 1977.
Moss, D. "Medard Boss and Daseinsanalysis." In *Existential Phenomenological Alternatives for Psychology*. R. Valle and M. King (Eds.). N.Y.: Oxford University Press, 1978.
———. "Phenomenology and neuropsychology: Two approaches to consciousness." In *Metaphors of Consciousness*. R. Valle and R. von Eckhartsberg (Eds.). N.Y.: Plenum, 1981 a.
———. "Erwin Straus and the problem of individuality." *Human studies*, 1981b, 4 (1), 49–65.
Spiegelberg, H. *Phenomenology in Psychology and Psychiatry*. Evanston: Northwestern University Press, 1972.
Straus, E. "Wesen und vorgang der suggestion." *Monatschrift für Psychiatrie und Neurologie*, 1925.
———. "Das problem der individualitat." *Die Biologie der Person: Ein Handbuch der Allgemeinen und Speziellen Konstitutionslehre*. vol. I. Berlin and Vienna: Urban and Schwarzenberg, 1926.
———. *Geschehnis und Erlebnis*. Berlin: Springer, 1930.
———. "Education in a time of crisis." *Black Mountain College Bulletin*, 1941, 7.
———. "The upright posture." *Psychiatric Quarterly*, 1952, 26, 529–561. (1948 a)
———. "On obsession: A clinical and methodological study." N.Y.: *Nervous and Mental Disease Monographs*, 1948 b.
———. "The sigh: An introduction to a theory of expression." *Tijdschrift voor Philosophie*, 1952, 14 (4), 1–22.
———. "Man, a questioning being." *Tijdschrift voor Philosophie*, 1955, 17 (1), 3–29. (1953)
———. "Der Archimedische punkt." *Tirage a part de Recontre/Encounter/ Begegnung. Contributions a une psychologie humaine dediees au Professeur F. J. J. Buytendijk*. Utrecht/Antwerp: Uitgeverij het Spectrum, 1957.
———. "Die verwechselung von reiz und objekt." *Verhalten und Verstehen*, 1963 a, II, 4–32.
———. "Born to see, bound to behold." *Tijdschrift voor Philosophie*, 1965, 27 (4), 659–688. (1963 b)

————. "Psychiatry and philosophy." In Natanson, M.(Ed.): *Psychiatry and Philosophy*. N.Y.: Springer, 1969. (1963 c)

————. "An existential approach to time." *Annals of the New York Academy of Sciences*, 1967, 138 (2), 759–766.

————. "The monads have windows." In *Phenomenological Perspectives: Essays in Honor of Herbert Spiegelberg*. The Hague: Martinus Nijhof, 1975.

PART ONE

Event and Experience (1930)

Foreword to the 1930 Edition

An investigation such as the following, which begins with the facts of compensation-neurosis and returns to them, must in the present condition of psychological and psychopathological research necessarily take up the discussion of a series of general questions. Certainly were our sole intent to intervene in the conflict of opinions concerning compensation-neurosis, the pathway taken here would be too wide and too laborious. But compensation-neurosis is not the final goal of these investigations; it serves rather as the medium—the optic system, as it were—to render visible the relationships currently existing between empirical, individual research and theoretical psychology. Our theoretical discussions thus also lay claim to an independent interest. In spite of the diverse nature of their objects, they all cluster around one problem—the problem of time. We have not yet attempted a systematic, comprehensive treatment of this theme; our aim is merely to outline the significance of time for the organization of phenomena in several areas of mental experience. Even from this selective presentation, however, we may conclude that time is in fact the central problem, or the axis, of theoretical psychology, around which all other problems must be organized.

Historical formation extends all the way down to perception; in the investigations presented here we will show how much the resolution of concrete psychological problems depends on the discovery of their historically relational nature. However, if perception itself can be entirely understood as merely an event, as a process, then it cannot also serve—as has till now almost uncontestedly been the case—as the foundation of empirical psychology. Other mental phenomena must become the departure point for psychological research. Memory is primary among such phenomena. But the theory of memory must allot the proper place to the concept of the New. Until now philosophy and psychology have almost completely misunderstood the problem and the breadth presented by the concept of the New. Only such a revised theory of memory will make it possible to comprehend mental being as an individual historical continuum and will allow us to give a chronological definition of psychic events in place of the chronistic.

Berlin, March 1930
E. STRAUS

Foreword to the 1978 German Reprint Edition

Why nearly five decades after the publication of this work a new edition? Is the scientific progress of psychiatry, of psychoanalysis, and of psychology not so general and beyond doubt that the critical position taken here by Erwin Straus can be shelved with the documents of the medical historian? Have the theses posited here not been developed further by numerous more recent investigations, and long ago been taken up into the edifice of general theory? And are the psychic trauma and the compensation neurosis, the themes specified in the subtitle, not somewhat antiquated and long clarified concepts, without great topical and practical significance for our time?

Erwin Straus poses a fundamental question here, when he pursues the relationship between events of the external, physical world—the world as a medium for us—and inner experiencing. What the philosopher treats as a knowledge-theoretical problem becomes for the physician a highly practical question. It is his responsibility to recognize the illness-inducing influences on mental existence, and to treat them preventively. He must ask: How is the psychic reality of the human being constituted, in its historical becoming, under the influences of the environment, and under the conditions of corporeal existence?

The continuous train of thought in this small multi-faceted work is the intent, never to lose sight of the lawfulness of the life-historical flow of experience, and to defend this lawfulness against the thinking and the conceptual system of the objectifying scientific approach. The physician knows directly that the mode of experiencing of human existence is relatively dependent on both external influences and the sustaining functions of the organism. The physician has much less access to, and typically observes only as a complication, the ways in

5

which the human being organizes this environment in his perception and in his inner experience-history. The historical, i.e., the life-historical structuring of every perception, every image, and every recollection should be conceived within their own appropriate order.

But is that not self evident in the psychology, the physiology, and the psychoanalysis of our time?

Certainly psychology, through the phenomenological research orientation, has come to recognize the subjective co-ordination of perception, of memory, and of motivation. However, psychology has isolated this particular form of questioning from its current research and applied practice, which aim at the measurement of behavior, or in the case of therapeutic practice at the modification of behavior, and do not aim at understanding, introspection and insight. Psychophysiology and psychosomatics recognize that psychic stress exists, for example when they describe the effect of environmental influences on the body, under the modern concept of stress. The effect of this stress, however, operates as an "inner stimulus" governed entirely by the stimulus-reaction model and according to the paradigm of physics. And does psychoanalysis, when it speaks today of the psychic apparatus and of anxiety as the sum of excitations, really use only an antiquated form of speech, which has been filled with an entirely different set of contents? For Erwin Straus, in any case, it was never indifferent, in which language and conceptual system the processes of experiencing were comprehended. He was convinced that only the most suitable language can maintain the connection between theory and the domain of things. He attacked the tendency to utilize vague, ambiguous concepts, and to think in analogies or in accommodation to the language and thinking of the natural sciences, whenever the understanding of subjective experience became difficult.

The situation within which psychoanalysis even today thematizes the environmental influences affecting the human being, occurs entirely according to a physical paradigm and an interaction theory modeled on it. Either the influences of the parents, and the child's introjection of them, are described entirely according to a bodily mode of operation, or else the analogous model of imprinting is utilized—a model first introduced in the behavioral investigation of animals. Even though psychoanalysis may speak less frequently of the psychic trauma than in the early days, nevertheless, in the fundamental work by Laplanche and Pontalis, this concept is still presented entirely in the classical Freudian sense of an excessive flood of stimulation to the psychic apparatus.[1] This demonstrates that the trauma concept, like many other concepts in psychoanalysis, and like the stress-concept in psychosomatic medicine, has its roots in a physicalistic thinking, which continues into the present.

The concept, introduced here by Erwin Straus, of the representative experiences, which, in a free or forced ''sense-derivation'', are preserved in memory by the individual, offers psychoanalysis more than a theoretical keyword. More importantly, it can defend psychoanalysis against false causal thinking in psycho-genesis, and from an elimination of the subject.

In 1930 psychoanalysis failed to recognize that this work challenges it to reflect on its psychological foundation, namely that mental existence be studied as an individual historical continuum.[2]

Ludwig Binswanger, drawing on the dialogue with Erwin Straus, and with reference to him, has characterized psychoanalysis as a hermeneutic of the historical stream of experience, which he called the inner life history. When, four decades later, a new methodological discussion commenced with Jürgen Habermas, this critical discussion had still not yet made a mark on the theoretical understanding of psychoanalysis, as one sees, for example, in the present day practices of psychoanalytic training. Erwin Straus proposed to take a science of the life-historical experience of subjectivity as the theoretical foundation for psychoanalysis. This proposal has influenced psychoanalysis, up until the present day, no more than Habermas' findings about the scientistic self-misunderstanding of its metapsychological statements. Habermas has shown that it is impossible to conceive neurotic symptom formations and early childhood development within the energy-distribution model.[3] And the proposal of Erwin Straus, for an historiological foundation for psychological theory, emphasizing the subject centered experience of sense-derivation, against the backdrop of a general theory of interpersonal experiences, is still waiting to be taken up.

One might ask how it was possible that psychoanalysis, in spite of its methodological weaknesses and its abbreviated theoretical self-understanding, has nevertheless been so fruitful in its practice, and has also steadily expanded the area of its observations, findings, and applications. Obviously one can make successful interventions in interpersonal understanding and behavior, without coming to a fully

1. Laplanche, J., & Pontalis, J.B.: *The language of psychoanalysis*. London: Hogarth Press, 1973.

transparent theoretical self-understanding. In the praxis of therapeutic dialogue, in the introspective and communicative access to the human being, and in the findings about the inner conflicts undergone by a man in the course of his developmental history, there is inherently an element which—at least in practice—sustains and further fructifies them. It remains an open question, however, whether this element can permanently, and without suffering further injury to itself, endure in a strange language and in categories of thought bound to the natural sciences.

This book by Erwin Straus breaths the spiritual fullness and arresting freshness in thinking of this multi-faceted man, then in his fourth decade. It is less systematic and, for the readers who are not psychiatrists, more easily accessible than his great work, appearing three years later, *Vom Sinn der Sinne* (English version: *The Primary World of the Senses*). The later work carried further the debate with Pavlov, begun already in the 5th chapter of the present volume. But already the discourses of *Event and Experience* presuppose the same universal education which the author, as neurologist, psychiatrist, psychologist, and philosopher had. In our age of specialization, it might today be difficult to find its equivalent among prospective readers. We believe, however, even though many passages seem to stray quite far afield, that this work can be a fruitful one for many different disciplines, insofar as it compels one to a reflection on methodological questions. The psychoanalyst will find, in this work, rich points of departure for a critical confrontation regarding the understanding of the neuroses. Erwin Straus deals here not only with the traumatic neurosis, rather he also makes several statements about the phobic neurosis, which were then later adopted by von Gebsattel and developed further.[4] Further, Straus gives examples of existential neuroses, in which the symptom is

2. Schultz-Hencke and Gerö granted some acknowledgement to the work, in their discussion of it at that time, but they believed that Straus lacked the empirical presuppositions of psychoanalysis, leading him into misunderstandings. All the more remarkable in this regard was the reception of this book among Straus' spiritual fellow travellers within psychiatry. Ludwig Binswanger devoted an essay to this work ("Geschehnis u. Erlebnis", *Mschr. Psych. Neur.*, 80, 1931). Viktor Freiherr von Gebsattel found many inspirations here for his works toward an anthropology of the neuroses and of the sexual perversions. ("Süchtiges Verhalten im Gebiet sexueller Verirrungen."*Mschr. Psych. Neur.*, 82, 8, 1932). Alfred Storch and Hans Kunz have referred, in their discussions, to the significance of this work. (*Der Nervenarzt*, 3, 607, 1930. *Z. ges Neurol. Psychiatr.*, 59, 7, 1931).
3. Habermas, J.: *Erkenntnis und Interesse*. Frankfurt a.M.: Suhrkamp, 1968.
4. Gebsattel, V. von: *Die phobische Fehlhaltung*. In: *Handbuch der Neurosenlehre und Psychotherapie* II. Munich: Urban & Schwarzenberg, 1959.

representative for the successful or miscarried life-project. Certainly Erwin Straus has not thereby dealt with the neuroses as a whole, nor rendered the psychoanalytic approaches to understanding superfluous. The actuality of the experiencing, into which the psychoanalytic assertions lead, reaches much farther than the evidence of what is directly visible would lead us to suspect.

Erwin Straus was more sympathetic with the luminous surface than with the deep-lying passions and the human instinctual world, more at home with the real conflicts and existential wrong-attitudes than with the sexual and aggressive motives arising from the archaic depths. The Apollonian was much closer for him than the Dionysian; the poetic word meant more to him than the obscurity of the mythos or even the difficult to comprehend declarations regarding the mythology of instincts, which characterize the work of Sigmund Freud. Thus many objections can be formulated from the standpoint of psychoanalysis. When Erwin Straus, in his case history of the 17 year old with the anxiety neurosis, describes this patient as "a good-natured, trusting, and candid youth with an open, free look, without any tendency to histrionics", and determines that not the slightest evidence could be found for guilt feelings in him—i.e., guilt over the unconscious death wishes—awakened by the real death—toward the uncle and the father, it nevertheless remains questionable, whether a psychoanalyst would not have opened up an entirely different perspective of dark, anxiety provoking phantasies and perhaps even another therapeutic prospect. The psychoanalyst will also establish that not only foods joyously consumed, but rather also foods intensively desired in shameful phantasies, can call forth guilt feelings and remorse.

Erwin Straus established a value-assessment for actions, both for actions on one's own and for those toward others, and he characterized the development-negating fixation to regressive needs as a "deformation". This concept became, for a time, a signpost for the understanding of sexual abberations, but also met considerable criticism.[5] Erwin Straus asserted of psychoanalysis, that it was in danger of viewing the child too much from the perspective of the adult. The application of the same concept, sadism, for both the behavior of the child and the sexual practices of adults, must dispose phenomenologists toward criticism, even if the two phenomena are connected in a developmental-historical

5. Gebsattel, V. von: *Süchtiges Verhalten im Gebiet sexueller Verirrungen. Mschr. Psychiatr.*, 82, 8, 1932.

sense. But is it actually true that there are no negative and destructive valuations in the child? In the development of the child, are not standing and going, or grasping and scratching, or tearing and cutting up, already behaviors which can succeed or mis-carry, or in which a reactive-destructive attitude may develop directed against the fundamental norm immanent to the thing? One of the merits of this book is that it forces the psychoanalytic reader directly to this reflection as well as to critical counter-questions. Erwin Straus sees the strength of psychoanalysis in its rootedness in the biological, i.e., in this case in embodiment as the foundation for the selection of representative experiences and self-consummating sense-derivations. That embodiment has a function as vehicle, but is also always historically mediated, must be viewed as a fundamental discovery of psychosomatic medicine.[6]

This new edition of this book goes to press with the hope that those addressed by the work, the psychoanalysts and psychologists, will give more heed to the statements of the author than they did five decades ago. Psychoanalysis has, in its theories, since then become ever more estranged from the historical modality of experiencing. This extends even to the proposition of David Rapaport, who, in what is by far the most widely disseminated work on psychoanalytic theory, set forth the thesis: "The object of psychoanalysis is behavior." It is a great misunderstanding, when even psychoanalysis itself places the accent not on its introspective potential, but rather on the observable and do-able, and when it would choose the behaviors to be altered as its central focus. The process commencing under psychoanalytic understanding does not permit any concrete behavioral directions, and also intends no direct modifications of behavior. The psycho-therapeutic transformation consists in the fact that "a past event has gained a different meaning for the experiencing individual;" this definition put forth by Erwin Straus himself, in the final lines of this book, is axiomatic for psychoanalysis as well.

Heidelberg, March 1978

WALTER BRÄUTIGAM
(*Translated by Donald Moss*)

---

6. See also Ludwig Binswanger: *V. von Gebsattel. J.f. Psychol. und Psychother.*, 6, 305, 1959.

1 / *The Representative Function*

In the newly reignited dispute concerning compensation-neurosis, the principal concern is the resolution of specific clinical and forensic-medical questions. This suggests that we must establish, by the usual inductive procedure and through the observation of a single case or of many cases, exactly what predicative terms define necessarily and unequivocally compensation-neurosis. These terms would be derived from the conceptual system of pathology or psychopathology, whereby we presuppose that conceptual systems themselves are already entirely or at least extensively defined. In almost all recent publications, however, an entirely different tendency has prevailed, namely, the tendency to seek definitions for these terms. Besides the attempts to penetrate further into the life-historical foundation of compensation-neurosis as well as into the structure of the social factors participating in its constitution, we find discussion concerning the concept of sickness, the concept of causality, and the concept of probability. In other words, it has become evident that those terms that should serve for a definition of compensation-neurosis, are themselves of the most highly problematic nature, and therefore, that the investigations aiming at an inductively based definition of compensation-neurosis, must be preceded by other investigations which, independently of these, submit the nature of the terms and concepts used for definition to a clarifying analysis.

Methodologically these are two entirely different tasks, which in many publications have not been kept sufficiently distinct. Thus we cannot decide the question—whether or not the compensation-neurotic can rightly be characterized as sick—if we have not previously resolved the more general question—whether the concept of "sickness" has the same meaning and validity in the regions of somatic medicine,

11

of the psychoses, and of the neuroses or whether we are merely playing a deceiving game with our equivocations. In turn, this question can be decided only after a methodical investigation of the nature of neurosis. Thus the difficulties that arise as we attempt to determine the nature of compensation-neurosis once again point out how indispensable general theoretical psychology has become for the solution of empirical clinical questions. The methodological difference between the two tasks is also not annulled by the fact that these very cases of compensation-neurotic illness can perhaps function clearly and representatively in our reflection on these problematic general terms and concepts.

Such a scientific situation—one that calls upon or requires an empirical problem from a separate discipline for general theoretical investigations—is nothing unusual or strange. Only someone imprisoned in positivistic views, who overlooks the theoretical content even of mere perception, will consider such investigations as superfluous and unfruitful. Every act of observing, even in the simplest form, begins with a question. Things answer only if they are questioned, and they answer only to that about which they are questioned. Certainly the scientist, insofar as he is not a psychologist or a theorist of epistemology, is not interested in the dialectical process of questioning and answering, in which consciousness unfolds as a becoming-conscious; what interests him is the answers, which amalgamate themselves into an always unfinished system of knowledge. However, whenever our discussion concerns not the knowledge but the learning, the gaining of the knowledge, we cannot overlook the fact that in their emergence the answers are systematically conditioned doubly through their indissoluble bond to the question. One cannot question arbitrarily but only from a definite theoretical standpoint.[1] This is even more generally valid for everyday, prescientific experience. The historical place for questions—and the place for answers—depends, on the one hand, on the unfolding of a system of objects and methods that affect and permeate one another in the most intimate fashion and, on the other hand, on the system of the individual person of the questioner. The overlapping of the two systems renders more difficult the knowledge of the lawfulness of the historical succession of questions, just as a curve that has arisen out of the overlapping of two different kinds of curves

---

1. We must strictly distinguish between the questioning, which reaches forth into the future, and the seeking, which is merely driven.

requires a more painstaking analysis. The mutual interference of the systems conceals—both for the verifying experience of everyday life and for the knowledge-expanding activity of the scientific researcher—that which is immediately evident in the learning or in the appropriation of a familiar topic of knowledge. It is self-evident to us that learning, even when it does not coincide with a process of maturation, must adhere to a definite course, for example, one cannot learn differential calculus before learning one-times-one. The course of instruction follows the inner order of the object. Discussions concerning practical pedagogical methods ultimately center on the question of whether the system of the object or that of the person should be brought more into recognition. In the same way, research must follow the system of the object—which must already be defined and ordered theoretically—if it is supposed to be able to be observed (that is, questioned) further at all.

Thus even "pure empiricism" is necessarily theory, although the pure empiricist often directs himself by theoretically unclarified and vague concepts. The distinction between empirical and theoretical investigation is in general not an absolute one; compensation-neurosis is an example of how much the so-called empirical object is determined by general concepts. The current difficulties derive ultimately from the inadequacy of these general definitions, which are implicitly involved in every case. Frequently in these general discussions, one hears the expression "*mere* theorizing," usually meant in a critical sense. Such a criticism would be justified only if someone without clinical experience with compensation-neurotics attempted a definition of compensation-neurosis from the standpoint of his writing table; and even then the use of the term "theory" or even merely "speculation" would be false. Actually such a procedure could at best be called methodologically indefensible phantasizing.

In contrast to the discussions during World War I, a step forward and general accord have now been achieved insofar as no one any longer doubts the mental conditioning of compensation-neurosis. Occasionally someone will proclaim that new investigative methods (for example, encephalography) have again provided evidence to justify Oppenheim's organic concept of traumatic neurosis; but such statements are based too manifestly on confusion between the question of diagnostic method and that of the nosological conceptions and cannot be taken seriously. Refinement of the diagnostic tool can show us only that we have classified cases incorrectly into the compensation-neuroses group when they actually belong to that of the organic consequences of accidents. By no means, however, does this impugn the basic distinction between these groups. Similarly, it would certainly not invalidate the concept of the organic consequences of accidents should an improvement in investigative methods successfully unmask as malingerers a few of the patients we had earlier viewed as organically injured.

Corresponding to the concept of the psychic origin of compensation-neurosis, the concept of the accident experience has gained importance. The expression "a reaction to the accident-experience" is frequently applied in expert opinions. In its present form, however, this expression is still ambiguous and misleading and therefore, in the absence of further clarification, is not suitable to be included in texts of law, as has been proposed. Thus, to emphasize only the two most practically important meanings, this expression—reaction to an accident-experience—could mean, on the one hand, those alterations taking place in and with this experiencing, that is, alterations of the experience-world and of the behavioral modes of the person concerned (which in turn are dependent on the experience-world) and, on the other hand, typically histrionic compensation-neurotic symptom formation in the narrow sense. These are clearly two entirely different modes of experience, so different that the word "accident" doesn't even have the same meaning in the two groups. The first case involved "*this* accident" as a one-time concrete incident, in which one personally participated, but the second involves the ideal supposition: an "accident," in other words, an image endowed with definite legal claims.

Let us assume that someone has taken an ample portion of an effective sedative prior to a night journey in a sleeping car; as a result he sleeps unusually deeply and soundly. During the night a mishap occurs, but the passengers of the sleeping car nevertheless come through without difficulty; they are of course soundly shaken up and in confusion, but the car remains undamaged, and they are able to continue the journey to their destination. Only the next morning does the sleeper, who has suffered only a few abrasions, hear about what took place during the night. Under the lingering effect of the narcotic agent he still does not experience fright, even during the narration of the event; rather he feels himself pleasantly safe. He has thus not experienced this accident at all. If in spite of this he should later, as a compensation-neurotic, come in for an expert opinion—and the example has by no means been fabricated out of thin air—then one is dealing not with the reaction to the *experience* of the accident but with a reaction to the *fact* of the accident. If in the one case it is the living through of this accident-event that determines the further experiencing, then in the other case it is the knowledge of having been present in an accident. Clearly the legal definitions concerning the compensation for an accident can intend only one of these two meanings; it is also clear—and through the emphasis on the original context also certain—which of the two is intended. Only this *hic et nunc* given,[2] in its peculiar nature, could be described as an effective cause.[3] It therefore remains to be investigated, what mental alterations the accident—this definite one-time event—could bring about in the person concerned.

In order to be able to distinguish terminologically the different types of mental relationship to the accident, we wish, for the category in question, to speak of the psychic trauma that an accident brings about rather than of a reaction to an accident-experience. From an analysis of this concept we may expect a clarification of the state of affairs in the dispute concerning compensation-neurosis because it can for the first time provide us with a knowledge of those psychically conditioned consequences of accidents, which could be compensable according to legal definition.

An exhaustive analysis of the concept of *psychic trauma* is at present all the more necessary because it has—in common with the concept of accident-experience and other expressions taken from everyday language or popular psychology—an ambiguity and indefiniteness that make it altogether suitable for common and casual usage but not for scientific knowledge or jurisprudence.

We would certainly understand each other with ease if we were to describe as psychic trauma, a threat to our bodily, economic, or social existence, if we were to describe as psychic trauma the sudden death of someone close to us, or if, as psychoanalysis in particular did in its first years, we were to describe as psychic trauma a sexual attack. But, since it diverts us from rigorous analysis, it is precisely this easy possibility for agreement concerning such an expression that could prove to be a hindrance in the advance of scientific research. In spite of the extensive

2. *Givens*: Assumptions, independent variables, or previously existing conditions.
3. In this matter we still leave undecided to what extent the concept of causality is applicable to psychic relationships in general.

usage that Freud made of this concept, especially during the first period of his work, he and his disciples have even now done very little to penetrate more deeply into its nature. Even in the work of his old age, "Inhibitions, Symptoms, and Anxiety," he still resorts to the same summary explanation he had already used earlier: that in psychic trauma such a quantity of excitations are conveyed to the mental apparatus within a short time that the apparatus fails to master them, thus producing lingering disturbances in the distribution of energy. Even if we wanted to ignore all the deficiencies in the details of this explanation, a cardinal error remains deeply rooted in the Freudian conceptual system, that is, the attempt to interpret the meaningful content of experiences as an alteration in biological functions. In place of the transformation of the experience-horizon brought about in traumatic experiencing, Freud substitutes the altered condition of the psychophysical apparatus.

The reproach for having—at difficult but decisive points in psychological concept formation—taken refuge in biological explanations, certainly doesn't concern Freud alone; it must still be extended to the remaining "biologically based" psychological theories. At point after point in psychology and psychopathology one encounters a behavior of the researcher comparable to that of the diving-duck. Just as this duck at every sign of approaching danger quickly disappears under the water, so too the psychologists, facing newly emergent psychological problems, gladly seek refuge under the surface of biology. The reasons for this are not difficult to discover. No one today wants to be regarded any longer as an associationist-psychologist, but the general human need to speak in images instead of in concepts had found its full satisfaction in the psychology of associationism. That need requires a replacement, and biology provides it. In my investigations of suggestion I ran into a state of affairs entirely analogous to that of psychic trauma. There, too, it was customary to explain any change in intentional givens as alterations of functions. All such explanations have the advantage over psychological analyses of relative simplicity and vividness because they avoid the true problems. Thus they are quicker to find a receptive listener, even though they are false. Purely psychological investigations, on the other hand, have more difficulty procuring a hearing. They are more complicated and deviate much more from the natural prescientific concepts and manner of thinking than do biological images. Biological explanations resolve psychological problems mostly through familiarity rather than through knowledge, that is, they classify them as an instance of one of the classes long familiar to us. But what is familiar still need not be understood for a long time. Thus anyone who hopes to gain attention for psychological analysis must first prove the insufficiency of biological explanations; he must show that they leave problems standing unresolved. Since the following investigation of psychic trauma must also take as complicated a shape as the facts of the matter now require, I feel compelled first to declare my position concerning biological interpretations. If in doing so I direct my polemics particularly against psychoanalysis, this is not because its theory of biological interpretations of psychic trauma is the only offender but rather because it is the most prominent and systematically complete among the biological theories in this area.

Freud, as we have seen, explains psychic trauma by the economic principals governing the mental distribution of energy. In accordance with this principle, ultimately, what is characteristic for psychic trauma is not the peculiar nature of the experience itself but rather its effect on the mental apparatus, that is, the disturbance in balance between the increase of excitations and motoric discharge. Thus Freud does not even attempt to determine more closely the mode of the traumatic experiencing; indeed it is sufficient for him to point to its theme: birth or castration, or their symbolic equivalents. A whole range of other experiences, however, could be interpreted according to this same schema of stimulus increases, for example, the effervescence of love or the moment of creative inspiration. How do such diverse experiences differ from psychic trauma in the way individual historical moments are lived? How can we distinguish them? There are manifestly too many different kinds of experiences potentially covered by the same biological metaphors; psychological theories of this kind have no explanatory value whatsoever. Stimulus, stimulus-increment, quantity of excitation, discharge—these trusted expressions are familiar to us all; for many they are also seductive because of their seeming obviousness. But as soon as we push further to learn more precisely what stimulus, quantity of excitation, and mastery are supposed to signify in an individual case, the insufficiency of these expressions becomes evident. Can one bring about the removal of the quantity of excitation through any arbitrary kind of motoric discharge? Cannot the mere correction of an effectively traumatic piece of information already bring about the disappearance of the so-called quantity of excitation?[4] Is there any quantitative relationship whatsoever between the "stimulus-increment" and the measurable portion of an action that serves to master the stimulus-increment? In a dramatically critical situation cannot a single word—"yes" or "no"—already break up the tension? No doubt can remain that after an upsetting experience a man does not recover his equilibrium by means of a simple process of motoric discharge; rather his movements must have the meaning of an action that establishes a new and different order.

Blind rage can be a substitute for action. In such a case, however, it is not that an unordered series of movement-discharges that have greater energy delivery takes the place of an ordered series with less consumption of energy; rather an action, which in a manner distant from reality symbolizes its meaningful import, takes the place of an action that realizes its meaningful import in a manner adequate to reality. The directionlessness necessitates a greater display of action on the part of the individual in a rage, and it must renew itself continually because it never reaches its goal. Exhaustion finally fixes a limit to raging; the individual in a rage stops "when he can no longer go on," and not because he has finally mastered the stimulus-increment. After an intermission for recovery he can thus begin his fury anew since it is not the measurable energy but the sense of what is done that signifies the dispatch or mastery of tasks. This sense, however, belongs to another domain than do the biological processes that actualize it. While these processes begin at a definite point in time, end in another, and pass away with their occurrence, the sense of the action does not become the booty of the past in the same fashion.

In the Freudian concept of psychic trauma, the already vanquished theory of traumatic neurosis lives on in altered form, with this distinction: in the place of Oppenheim's hypothesis of anatomical molecular alterations we find a functional pseudoenergetic view.

Here a psychoanalyst might object: that is one of the customary misunderstandings; in fact we distinguish expressly between psychic energy and its physical forms. We also deny that any simple relationship exists between intensity of feeling and the quantities of energy

---

4. The fact that an improving, a correcting, or a rectifying exists at all is important for psychological theory formation. However, like so many other relevant phenomena, they are almost never considered and appreciated in their proper significance.

carried off. Otherwise we would in fact "leave those energy displacements taking place completely in the Unconscious entirely out of consideration" (Hartmann). The objection is literally correct, but energy that is not measurable and in fact is not measurable in principle—not merely because the present state of methodological technique does not yet allow the measuring—is a scientifically worthless and misleading concept. With the emphasis on the impossibility, in principle, of measurement, we have an admission that the attempt to base psychology on a biology that is regulated quantitatively according to the pleasure principle cannot be carried out. The natural scientific concept of energy has a place only within the natural scientific time-system. Whoever applies this energy concept to psychology mistakes the time-structure of experience, which is of an entirely different kind and is thereby forced to build up the life-history summatively out of individual moments. The dilemma resulting from the attempt to bring his psychological insights into harmony with this inadequate biological theory compelled Freud to continuous contradictions in his points of view. He never succeeded in unifying them. The energetic view banished him into an atomizing psychology. Ultimately this obstructs access to all experiences bound to a definite meaning structure and historical modality, such as trauma.

The energetic view remains necessarily bound to the stimulus-response schema customary in the psychology of the senses. The energetic point of view thus refers back again and again to sensations and complexes of sensations as the ultimate independent constituents of experiencing. Many parts of the psychoanalytic system could illustrate this point. We are only interested here in the fact that Freud again and again conceives perception and the image as particular, isolated forms originally constructed out of the contents of sensation and as likenesses of the sensually-evident givens, whose after-effect can be explained only as the continuance of particular images in memory or in the Unconscious. The image goes into the Unconscious just as the particular act of perception has stamped it. With the concept of the special mode of functioning of the system *Ucs*[5], Freud makes an effort after the fact and by means of a material hypostatization to bring in

5. *Ucs*: the Freudian abbreviation for the systematic Unconscious, that is, the psychic system, which contains all contents and processes that have actually undergone repression. This system is characterized as functioning in a manner violating the logical, continuous, and consecutive order of our waking life (Translator's note).

along with the image the manifold relations, within which the individual image is necessarily always situated. Our further statements concerning psychic trauma will show, however, that the assumption of isolated perceptions and images in no case, not even in the most primitive givens, does justice to the character of the experiences. Certainly in a perception I intend "this table" here before me as an isolated, relatively independent thing; nothing is more injurious to psychological knowledge, however, than to misinterpret the properties assigned to the perceptual object as properties of the act of experiencing, perceiving, and imagining. This holds true even for such a simple object as the perception of a table or of a colored surface but even more so for richer experiences such as psychic trauma.

No matter what hypotheses one may contrive to supplement the energetic theory of the mind, a mental energy could nevertheless be meaningfully attributed in general only to the psychophysical apparatus in its changing conditions. Only the contents of sensations, however, and never the intentional objects, can belong to the conditions of the psychophysical apparatus. Thus in the terminology of modern psychology only the contents belong, but not the objects.

Psychoanalysis does not overcome these difficulties; rather it goes around them in that it makes use of ambiguous concepts which are overburdened with evident import. Freud employed expressions such as "the drawing back of libido from objects," and "the return of object libido to the ego." At best these are similes, which can illustrate mental occurrences through concrete processes, but such metaphors do not have a real knowledge value. Already with the term "object libido" we discover that it is based on a hybrid concept-formation, in which the physical, the physiological, and the psychological have been unified into a mixture that is difficult to break down. Psychoanalysis emphasizes that libido is a quantitative, purely energetic concept, which is supposed to designate the psychic energy in analogy to physical energy. However, an energy that directs itself toward objects and which can return has only the same letters of the alphabet in common with the physicalistic concept of energy. Libido becomes, in the application used by Freud, animated material. The energetic point of view is made possible only by the fact that important features of mental life have been taken up from the beginning and included in this concept of energy, while at the same time this accomplishment is presented in such a way as though this were not the case. Characteristic for this is Schilder's position: "Everything that we do ultimately originates in our instinctiveness and its energies, but simultaneously there are also objective goals toward which we are oriented."[6] This thesis is, strictly speaking, not permissible within the frame of the psychoanalytic system; rather it negates the presuppositions of psychoanalysis. It is a concept that abandons the foundations of psychoanalysis, not in the question of secondary importance but at a decisive point. On the other hand it creates a difficulty, which psychoanalysis veils with the help of an unclear terminology. What is astonishing, certainly, is the ease with which Schilder skips over such a question of principle. His presentation does not settle the problem, but rather dismisses it. The entire problem hides in the "simultaneously." Psychoanalysis denies the simultaneity; for psychoanalysis the instincts and their energies come first while the orientation to the objective goals is a secondary manifestation dependent on them. If the alternative is set up as Schilder has formulated it, then one must decide for one of the two possibilities. Contrary to this, Schilder decides for both; he establishes the simultaneity but he does not solve the mystery of this simultaneity. He gives no synthesis in which a dissolution of the antitheses would be possible, but is satisfied with a side-by-side coexistence, with an eclectic unification of opposites.

It would never occur to me to deny the alterations of the bodily functions that occur with traumatic experiences.

If we had an apparatus that without disturbing the experience could register the total activity of the nervous system and measure processes of metabolism, it would surely provide us with characteristic curves. Meanwhile the most exact knowledge concerning the results of the serum-calcium test and other such procedures cannot teach us the least thing concerning the nature of mental trauma or other experiences.[7]

* * *

6. Paul Schilder, *Medizinische Psychologie*, p. 300; English ed., *Medical Psychology*, International Universities Press, 1953).

7. Today many fall prey to the prevalent illusion that psycho-galvanic experiments, chronaxy-determinations, plethysmography, and other investigations of bodily alterations produced by suggestion could contribute something to the resolution of the body-mind problem. In this illusion as well as in Prinzhorn's statements and other kindred statements concerning the body-mind-unity, we encounter the confusion of the metaphysical problem with the pragmatic one of psychophysical coordination. I have already referred to this distinction elsewhere. (*Monatschrift für Psychiatrie und Neurologie* [67], 1927.)

Following this broad but indispensable defense against a theory that obscures the problem, we can now finally begin with the progressive presentation of our own concept. We wish to proceed from an example. Let us assume that somewhere on a street someone has been run over by an automobile and killed. Among the group of those standing around the mangled body are a physician who has long since become indifferent toward this kind of impression and a youth who is presented unexpectedly for the first time with the sight of violent death. Quietly and in a businesslike manner the physician carries through his duty in this situation; through it all he remains inwardly uninvolved, and the experience has no further effect on him. The youth, on the other hand, cannot forget the sight of the corpse the whole week long. He is changed in his entire behavior: oppressed, quiet, anxious, and timid, he doesn't wish to go out on the street alone. From his original transient sensitivity he gradually develops a sensitivity heightened to the point that it becomes a defense against death and its manifold forms of aging, withering, fading, and poverty. And yet one thing must be mentioned: many sensory impressions present in the original experience—the quality of light, air, smells, and wind—have assumed a specifically repulsive character, which they maintain thereafter, even when detached from the original process, so that, emerging again some time or other in later years, they still exert an inexplicably strong effect distinguishing them from other similar stimuli.

How then does the original experience of the physician differ from that of the youth? With this question we are not asking about the variations in the later effect, because these are derived only from the variations in the original impression; they are only their temporal explication. Original impression and further behavior are not bound to one another energetically but rather through a meaning-connection. The not-being-able-to-come-free from an experience does not for the first time make it traumatic; rather, inversely, it is inherent in the nature of traumaic impression to force further experiencing in a definite direction. The external, sensual, perceivable process is certainly the same for both the physician and the youth; yet in spite of that the "stimulus-increment" is entirely different in importance for the two. It is thus not the sudden stimulus-increase that makes the experience into a traumatic one for the young man; instead the traumatic meaning of the experience brings about the so-called stimulus-increment. In other words, psychoanalysis employs the expressions of the stimulus-increment, the quantity of excitation, and the impeded motoric discharge, when in reality we are dealing with the meaningful impact, the structure and the mode of the experience.

In our example the familiarity of the observed process allows it to become a "particular case" for the doctor, almost in the sense assumed by sensualistic philosophy. The corpse lying before him is for him merely *some man or another*, a specimen of the species *homo sapiens*. His view does not penetrate to the concrete individual person whose existence was here annihilated. This individual person appears for the first time for his relatives and friends in their sympathy and grief. For the physician the following generality has once again been actualized through the accident: death and the destructibility of human existence. These general givens have long been familiar for him, just as he knows that it is inherent in the nature of death to occur at longer or shorter temporal intervals. Insofar as he thus experiences the process as a particular case, he subordinates it as the singular actualization of a general meaning; he also disposes of it in that he suspends his own moment of individuation by means of this classification. Since what is dealt with is a singular case, and since events of that kind take place with a certain temporal and locational distribution, the particular case is at the same time an isolated one. To this extent the facts of this instance do not alter the meaning of the generality that they bring up and therefore have no deeper reaching shock as a consequence. For the physician the entirety means no more than that here once again someone or other has met with an accident.

The experience takes an entirely different shape for the youth. His experience converges with that of the physician insofar as for him, too, the individual person of the accident victim plays almost no role (it is within our discretion to assume that the victim has been absolutely unknown to the youthful bystander). However, while the experience of the physician is described by the linguistic formula, "*a* human being" has met with an accident, the corresponding formula for the youth's experience reads "*the* human being" can die. For him the event has an entirely different representative significance than for the physician. For him what is disclosed in the particular process is death, mortality, frailty, and the vulnerability of the human being in general. As he experiences a new theme of universal meaning and existential importance, a sweeping transformation takes place in the youth's horizon of experience. Death has become conspicuously evident for him—a constantly lurking figure ready at any time, as it has been presented in the dances of death—and thereby he has also become aware of the vulnerability of his own personal existence.

It need hardly be said that the formulae "a human being has been killed" and "the human being can die" are not experienced, that rather the experiencing takes shape in conformity with these formulae. Language offers the possibility of expressing decisive distinctions between modes of experiencing through grammatical formulae because these formulae were originally formed with and from these modes of experience and have retained and concentrated these modes in themselves. The sense for the livingness of grammatical forms has been obscured through misuse carried on with the printed word. In poetic language and even in artistic prose grammar has preserved its original life. From artistic prose down to the prose of the journalist, language passes through a process of impoverishment and emptying-out of meaning. The judgment intended by the expression "prosaic"—a judgment carried over generally from language to modes of behavior—concerns only the lowest, last stage of prose, in which the language is entirely stunted and in which the symbolic fullness of the word has become a mere designation and the grammatical form a formula. At this point it can sometimes seem as though someone had put into the formulae a meaning-content originally strange to them, while in truth their very existence points to the meaning to be fulfilled in various kinds of experiences.

Only the need to allow the shocking experience to grow to its full intensity could lead the youth at this time to also turn toward the individual person of the victim and perform acts of empathy or of identification.[8]

Empathy with the individual person of the victim, however, is by no means necessary for the traumatic experience to attain to its own peculiar emergence and formation in the youth. Empathy can occur or not occur; in either case the experience again departs from the special circumstances; the individual person of the victim remains disre-

8. It is important for the theory of empathy as well as for that of identification in the analytic sense to point to the foundational relationships existing here. Empathy does not mediate understanding and knowledge of the other; rather, inversely, empathy is possible only where the individual already possesses the general meaning prior to the actualization of any particulars in an experience. The impossibility of achieving empathy is a diagnostic criterion to the extent that a common possession of the same general meanings and of the modes of behavior directed at these meanings can legitimately be presupposed for the average case.

garded, as are the time and locality of the occurrence as well as the details of the course of events and of the impression produced, for example, that the accident involved a man and not a woman, an adult and not a child, a well-dressed individual and not a poor man, and so on. The experience is instead more immediately oriented toward the general meaning of death given by the reality of the corpse. To this extent, for the physician too the experience must certainly advance from that which is clearly given to the general meanings founded in it so that what is sensually, obviously presented to him may be defined as such. Only to this extent can it become an object of meaningful perception. But while his experiencing immediately turns back again from this stage of general meanings to the particulars of the case—toward which his action must direct itself in its specifics—the traumatic experiencing ascends to further stages of general meanings.

They too are representatively contained in the particular event described in the example. The general meanings erect the pillars of the individual world edifice; the most general meanings stake out the space within which events generally can take place. The general meanings can be modified in their manner of manifestation—in youth more so than in old age; they continue to exist from transformation to transformation (each of which takes place in a shocking experience), unchanged in their content, and thus they determine what experiences are in general possible. The general meanings enter into what is objective but neither as functions nor as categories. They are bound to what is objective and are continuously presented to us from it, but we nevertheless do not, along with this, assume that they are henceforth persistently conscious. It is certainly highly conceivable that the youth of our example, in a report of his experience, would be able to tell only of the terrible impression, that in rough words he could describe the feelings that moved him and their occasion but would not know any more to say. The general meanings would be absorbed, as it were, by the picture that he had before his eyes; the historical modality of first-time-ness would fade into the feeling of shock. In spite of this, we advance the claim that the general meanings and the first-time-ness would be experienced exactly as we have depicted. Why he knows nothing and can say nothing of all this, what the conditions are for knowing and for linguistic formulation, and how that which is conscious relates to what is known, noticed, possessed, and so forth—to explain all this would require an investigation of a fundamental nature that we cannot carry out here. I believe, however, that even in the absence of references to the unconscious, our claim can hardly be unconditionally or absolutely contradicted. Thus we may register our claim here with the proviso that a later demonstration will offer proof.

The impossibility of linguistic formulation makes it difficult in certain cases for a third person (for example, the psychotherapist) to comprehend those general meanings determining the experiencing of a human being. If in urgent conversation or in psychoanalysis he encounters a shocking experience, then he too will likewise fall easily into the self-deception of the person reporting and mistake the representative portion for the whole of the experience. This deception has resulted in genetic theories.

Under the dominating influence of the English philosophers from Locke to John Stuart Mill and Spencer, psychology has long succumbed to this deception corresponding with the natural attitude, namely, that the obvious likenesses of the individual objects are the sole, true constituents of consciousness, and that concepts, especially general-concepts, are only names or signs for groups of individual images connected through similarity. Suffering under this deception, psychology could not recognize the relationship between the intuition or observation and the representative meaning and accordingly could not illuminate the structure of the experiences co-constituted through these relationships. For the practical man of action the final characteristics of things—especially their space-position and time-position—are important; they are the points at which his action can and must be initiated. His thinking therefore moves in a stratum of the most extensively specialized concepts possible—concepts that can be translated immediately into observation. This habit of thought leads to the deception, that the individual-concepts are mere likenesses of the objects.[9]

9. Even the Unconscious of psychoanalysis is for the most part filled with particular images that, through their timeless duration and temporally unaltered reality, function like general meanings. Freud reifies the general meaning; the naturalistic factor of the particular unconscious image, as a result of the special mode of functioning of the system Ucs, works just like a general meaning. Freud interprets the peculiar nature of the general concepts and their possible relationship to a manifold of individual objects to be the result of the primary process seizing upon the particular image (for example, by displacement and condensation). Freud overlooks that such a seizure is possible only because this "becoming-related" is inherent in the essence of the general concepts, that is, that this "seizure" absolutely could not take hold of the particular image as such. For him the particular image remains a particular image; it is also for him the self-evident element of psychology in general. It becomes evident here how much analytical empiricism is a product of the logical evolution of theory. Of course we must add that Freud

Living in the full accomplishment of his psychic functioning, the practical man of action never becomes aware that the individual-concepts, like the perceived objects themselves, are continuously co-constituted by the general meanings. His actual interest always glides away from this and toward that. First and foremost among all other attempts, Husserl's battle against psychologism and nominalism has served again to pave the way for psychological research in this area.[10]

Our new freedom, however, now places us again before new difficulties because, if the relationship between the intuition and the representative meaning exists in the manner we have claimed, then it must also be valid in general. Every object can become representative; it must be possible to ascend from each to a general meaning represented in it. Experience shows, however, that mental shocks—and among them psychic traumata—are rare, that the ascent to general meanings occurs only under certain circumstances, and, lastly, that even when it has occurred, a shock by no means takes place every time.

Thus our demonstration of the representative functioning of intuitions in the psychic trauma has merely opened for us an access to this problem. We are still quite far from any resolution to the analysis. We must ask which moments are included in the structure of the shocking experience, and which ones are merely the subjective and objective conditions for its occurrence. These questions must occupy us further in the following pages.

The next qualification concerns the general meaning itself. It makes a difference which general meanings one is dealing with. The logical form alone is not decisive. Of course it must first be a general meaning; it is not the particular threat itself but rather the experience of being threatened that generally suffices for a shock. Nevertheless, general meanings of a specific kind are necessary, namely, such meanings as concern the existence of the person. It will be sufficient here to refer to

---

remained ignorant of the problematic nature of these theories; fully naive in this, he took these theories over through the mediation of the natural sciences and experimental psychology of the nineteenth century. Freud jeered at the philosophers who do not wish to undertake their life-journey without a Baedeker to the metaphysical lands. But this battle against metaphysics cannot conceal the fact that Freud himself has neglected every confrontation with logic and theories of method, that is, every philosophical self-reflection and, in their theoretical portion, his teachings have thus become to a large extent a popular philosophy.

10. Cf. especially the chapter ''The Ideal Unity of the Species and the Newer Theories of Abstraction,'' in *Logical Investigations* II, Section I.

what is *existentiell*[11] as the thematic content of the shocking experience. In a later section we will discuss in detail this thematic content, the individual's readiness to receive it, his readiness to linger with and search for limit-situations,[12] and the opposite phenomenon—the individual's refusal of and flight from limit-situations—as well as the meaning,of these modes of behavior for the further organization of experience.

At this point, however, we must still consider the consequences of our declaring that only *existentiell* general meanings can produce a shocking experience. These consequences concern the connection between the representing intuition and the represented meaning. The general meanings, including those occurring in the psychic trauma, have an ideal value. Nevertheless, since they relate to existence it is necessary to endow these general meanings also with form (Gestalt) and presence in the sphere of sensually-evident reality; it is necessary to concretize such general meanings. (We are dealing here with a necessity originating from the total-constitution of the human being, and not with an essential connection). I believe I would best be able to illustrate what I mean here by presenting the history of a patient.

Shortly before I concluded writing the present work, I had the opportunity to observe a patient who impressively corroborated the analyses given here.

On the 23rd of January, 1930, the seventeen-year-old shopkeeper's apprentice, Martin M., came into my treatment. He had already suffered for approximately ten days from a condition of anxious, tense unrest, which at times intensified into anxiety attacks that the patient characterized as "indescribable." The first severe attack had occurred shortly after the burial of an uncle of the patient. During his convalescence after having had the grippe, at a time when he wanted to get up, this uncle had suddenly collapsed dead in

11. *Existentiell*: concerning the immediate existence of a particular, definite human being (Cf. M. Heidegger *Being and Time* [N.Y.: Harper & Row, 1962], p. 33). (Translator's note)

12. Karl Jaspers introduced to philosophy the idea of limit or boundary situations, *Grenzsituationen*. Limit situations such as death, suffering, war, and guilt serve to bring an individual existence into an encounter with its own limits. The individual thus experiences himself in such situations as unconditioned and reacts not with planning or rationality or fixed beliefs but with an intensified encounter with himself. Man lives for the most part in the form of a subject-object split and is absorbed with striving toward goals, purposes, and destinations in the world of objects. In the limit situation this taken-for-granted world of objects and the individual's orientation to it are undermined. Nothing is firm, nothing is absolutely beyond doubt, and one can get no "hold" on anything. The stage is thus set for man's return to himself. (K. Jaspers, *Psychologie der Weltanschauungen*, 5th ed. [Berlin: Springer Verlag, 1950], pp. 229–280). (Translator's note.)

bed. The patient had seen the corpse on the day after the death. The sight of the dead man, and particularly the suddenness of his dying, had made a deep impression on him. From that time he had to reflect a great deal over "being-dead": whether the dead still perceive anything, whether they are aware of anything about "being-dead down there underground." He stopped reading the newspaper because he feared happening upon the obituaries there. Above all he feared to encounter the dead and "saw spectres"—forms dressed in death shrouds that emerged next to or behind him and that even cropped up in his room, always in the periphery of his field of vision. He was fully convinced of the unreality of these manifestations, but this knowledge did not suffice to force the manifestations to disappear, or for him to chase them off or repulse them. In the severe anxiety attacks, which I was able to observe twice during his treatment, the patient could in no way be calmed by consolation or exhortation. He fell into a state of dyspnoea.[13] His forced breathing produced a tetanoid state of cramping in the arms and probably even in the breathing musculature. Hands and fingers fell into paw-like positions. Trousseau's sign, Chvostek's sign, and lowering of the threshold for electrical stimuli, however, were not demonstrable.[14] In the beginning of his suffering, on a walk with a friend in an ominously quiet but otherwise familiar street, the patient had an extremely tortuous feeling of alienation. The houses appeared to him as "ghostly," as if they were only walls with no life inside. Later the phenomenon of depersonalization occurred more in the bustling, crowded neighborhoods, while the quiet became pleasant for him. A pronounced unrest always befell the patient before going to sleep; the feeling of loneliness and abandonment was at that time especially strong.

One and a half years before the outbreak of this illness, the father of this patient had died. He had suffered from a *paralysis agitans* (idiopathic Parkinsonism) for the last six years of his life. The patient had to provide the larger part of the care for this man, who toward the end was almost completely helpless. He had gotten over this year of caring for his ill father as well as his father's death quite well.

In the course of the treatment the patient improved to the extent that the lingering anxious tension subsided and the severe, isolated anxiety attacks appeared only at infrequent intervals. The feeling of alienation, on the other hand, persisted.

Aside from the anxiety attacks no distinct alterations of the vegetative functions, metabolism, sleep, or nutrition were evident. During the attack-free periods the patient expressed himself without any motoric or linguistic inhibition; he complained occasionally of pressure in the head.

The patient is a good-natured, trusting, candid youth with an open, free

13. *Dyspnoea*: medical term for difficulty in breathing. (Translator's note.)

14. *Trousseau's sign* and *Chvostek's sign* are both medical diagnostic signs for tetany. *Tetany* is a disease in children and young adults marked by intermittent, painful tonic spasms of the muscles. It may be induced through hyperventilation. The test for a lowered threshhold to electrical stimuli is an *electromyogram*. (Translator's note.)

look, without any tendency toward histrionics. He is willing to work and has forced himself to resume his employment. Although his apprenticeship was not yet finished, he accepted a position of confidence in his firm. On the whole, in his demeanor he makes an impression somewhat more childlike than corresponds to his years.

An analytical treatment has not been feasible because of external reasons. We had repeated and thorough anamestic discussions, however, and these did not give the slightest basis for the presumption (which one could easily leap to) that, from the time of caring for his sick father, some feelings of guilt had accumulated in the patient and that these feelings had now come to a neurotic discharge. Active pubertal conflicts also did not appear to be present in the young man, who was not very strong in his drives, or at least such conflicts appeared to play no essential role.

We are thus dealing not with a conflict-neurosis but rather, if the expression is permitted, with an *existential-neurosis*. In the course of the treatment the impression had continually deepened for me that the delayed detachment of the patient from the family, as a result of the external circumstances of caring for the sick father, has been one of the conditions for the shock he has suffered. His development of independence commenced late and eruptively. He stood close before the transition from the sheltered and secure existence of childhood to the exposed existence of the adult. It is precisely for this reason that the vulnerability and solitude of his new form of existence—indeed of human existence in general—could become manifest to him in the sight of the dead man and in the experience of the sudden death, with a transformation of a catastrophic nature in his experience-horizon.

Regarding the symptomatology of this case, we are especially interested in how the general meanings of death and of vulnerability are transformed into reality. M. must ponder over the dead and over being-dead; at the same time, however, the entire environs become populated at point after point with dead men who terrify and threaten him from all sides. Of course he continuously sustains the critical perspective; the manifestations never win full reality, and he himself calls them spectres, but nevertheless the critical perspective cannot chase them away. They exist between reality and unreality. This is in fact the paradoxical mode of being of spectres, that they possess an unreal reality. Even for our patients the manifestations are only spectres. Yet as spectres they are nevertheless there; they have a place in reality. This is what I mean by the need to concretize the general meanings. In our case the general attains a *universal concretization*; the general becomes an everywhere.

This process of universal concretization looks exactly like an inversion of *Berkeley's* theory of abstraction. The concretization of the general also plays a large role in the thinking of primitives precisely because it restricts itself in content to what is *existentiell*. The mode of thinking of primitives described as a "participation mystique" is only possible through the concretization of the general. Finally, one could cite here yet another significant example: the dispute over images in the Western churches. The battle over the veneration of images in fact did not begin only during the time of the Reformation; on the contrary, it reaches far back into the early Christian centuries. The adversaries of images undertook this battle out of the fear that the veneration of images could become worship of the image. Translated into psychological language, they feared that the process of concretization could advance so far that the images would absorb all religious meaning substantially into themselves and limit it to themselves.

The need for the concretization of the general doesn't necessarily always lead to universal concretization. Besides this form, we frequently encounter in psychopathology the form of the *specific concretization*. It is present, for example, in those cases in which the development of a perversion can be traced back to a definite experience such as when a fetishist, in a definite situation, has experienced an erotic conquest for the first time, and ever thereafter remains bound to this specific situation or to a specific object—fur, boot, or similar item. Even in the first encounter the object has its effect only through the meaningful content represented in it, which could have been represented in the same way by a great number of other, different objects. Here, however, the experiencing turns immediately from the meaningful content back toward the object representing it, and remains entirely fixed to it, so that an inextricable entwining occurs between the representative and that which is represented, whereby the erotic object becomes absolutely specific. Much remains to be said in the course of our investigation concerning this complete entwining of representative and represented as well as concerning the process of the narrowing-in of choice, which underlies the being-fixed-to.

Before we conclude this section on the representative function, however, we must still consider one important factor of relevance here. Our examination of the need for concretization has already revealed to us that not just any arbitrary becoming-aware of general meanings is sufficient for the formation of the shocking experience. In the relation of representative and represented both members are indispensable. If the experience receives its mental fullness only through the general meanings, so too, only through the sensually-evident content provided by the representative, and through the space-position and time-position belonging to it, does it win the reality that lifts the shocking experience above and beyond mere theoretical knowledge about the general meanings. Under certain circumstances it can even serve as a soothing effect, to steer one's view away from what is individual toward general meanings. It is one of the techniques for consolation, both when I console another and when I discover consolation through my own surrender or resignation, to suspend the heaviness and bitterness of a personal privation by a glance at the general transitoriness of human existence. Thus here the general brings about security and assurance, in contrast to the effect of what is individual. The general can accomplish this, because it has already been situated beforehand within the whole of the personal world and has an acknowledged place in it.

A shock takes place only when, in the one-time experience, general *existentiell* meanings for the first time are presented to one's gaze and break into one's personal world. The shock thus depends on the historical modality, on the first-time-ness of what is experienced.

2 / *The Historical Modality*

In the example we took as our departure point an important distinction between the banal and the shocking experience was made prominent by our assumption that in the doctor we deal with a process already often experienced in a similar manner but in the youth with a first-time experience. That, however, was first of all only an objective determination of the first-time-ness and of the often-ness. As observers we determine that an experience of that kind has happened here for the first time and there repeatedly. However, if this distinction is supposed to characterize the experiences themselves, then the first-time-ness and the repetition must originally form the experience in a definite manner. Thus it is not only that, looking retrospectively from the experiencing individual, we establish or could establish that he already has or has not yet experienced something similar; rather, within the continuity of experiencing, the first-time-ness or the repetition gives the particular experience its special imprint. It is a question of the historical modality of the experiences, and this historical modality appears as a special moment in the experience itself.

Our investigation borders closely here on difficult problems of research on memory, in particular the question concerning the essence and origin of the quality of familiarity. Nevertheless, our formulation of the question can be brought into a sufficiently distinct contrast with that of the theory of memory that we can leave aside all questions of the theory of memory. We ask here neither how does the familiarity originate nor what is its basis in the particular experience; rather, we ask how first-time-ness and repetition constitute themselves in the historical passage of experiencing, and what they signify.

First-time-ness and transformation stand in a reciprocal relationship. With the first-time-ness the transformation comes to pass, but at the same time the first-time-ness is also founded in the transformation. That may sound strange, but it can be made intelligible without any far-reaching reflections. As is the case for the traumatic experiences under discussion, so too not all decisive experiences are repeatable. By no means do we need to go to the outer limits of experiencing. There repetition is often already excluded simply because the same situation cannot return. But even those meaningful occasions that are repeatable—an encounter, a journey, the reading of a book—are all the more disappointing in the repetition, the deeper the first impression has been. Disappointment even enters in when the merit of an art work or the beauty of a landscape is reexperienced. It is based on the fact that the transformation coming to pass in the first experiencing cannot come to pass again. Although those general meanings caught sight of in the first experiencing may show themselves again, they nevertheless show themselves only at that place in the experience-world to which they have been referred in the first experiencing. With their first emergence there occurred a profound metamorphosis of the experience-world in general. This metamorphosis has a temporal place in the historical continuity of experiencing from which it cannot be removed. The uniquely occurring metamorphosis of the experience-world is an enduring one, even though—as long as further metamorphoses are possible—it may not be final.

Thus the shock proceeding from meaningful events is connected not only to the content of the experience in the narrower sense, that is, not only to the fact that this and that appears in the visual field and is fully grasped in its representative meaning; but it is also connected to the historical modality of the incident and to the metamorphosis coming to pass with the appearance of the contents.

The subjective first-time-ness, however, need not harmonize with the objective. A situation can already often have presented itself in the course of a life before one's eyes finally for the first time open up to its representative general meanings. Often there are entirely individual nuances that make an object suitable to function as representative of a general meaning. Now for the first time the experience becomes meaningful; only now does the metamorphosis take place. The meaning-context into which the experience has moved now bestows upon it, and even upon its observable condition, a special character— the cachet of first-time-ness. There can be no doubt that experiences of this nature, in which a general meaning is disclosed not through progressive induction and abstraction but rather suddenly by an exemplary case, are not at all rare.

The descriptions of shocking experiences, particularly of religious conversions but also of their other forms, consistently emphasize not only the force with which the experiences seizes a man but also the suddenness with which it overtakes him. The "Sudden," however, is not the result of the rapidity of the changes in any arbitrary experiences whatever. If we look out onto the terrain immediately next to the tracks from the window of an express train traveling at full speed, one object after the other will spring into view in the most rapid succession. In spite of that, and even if the rapidity increases so much that the objects blur, we do not have the experience of the Sudden. The experience of the Sudden does not depend on the tempo of the external alterations; it depends entirely on whether or not amidst all of the change the continuity of meaning in the inner life-history remains preserved. Only if the basis is laid in the succession of objects for a rupture in the continuity of meaning do we have the experience of the Sudden. Even the suddenness of shocking experiences is not the result of external processes; on the contrary it is entirely the result of radical metamorphoses occurring in the shocking experience. The Sudden is the content of experiencing, not the external form to be related to objective time. It is not the Sudden that conditions the shock, but rather it is the extent of the transformation in the shock that conditions the experience of the Sudden.

Accordingly, we derive the enduring effect of the psychic trauma meaningfully from the metamorphosis of the experience-world occurring in the original experience, and we consider it fully incorrect to proceed inversely and to base the individual character of the psychic trauma on the intensity of the lingering effect.

Just as we later have at our disposal the general meanings originally acquired in a theoretical attitude, without any necessity to reproduce the experience in which we acquired them or the case from which we proceeded, similarly, the characteristic force of the traumatic experience does not rest on an unconscious continuing existence of the original manifest given. The limits of the experience in no case end, as it were, with the visual field. The so-called individual experience is always an event within the entire experience-world. This is not a mere juxtaposition or succession of formerly manifest givens; it is a system, built up in stages, of meanings and relations. The individual character of this system determines which individual experiences in general are possible. Thus the transformation of the experience-world, which occurs once and for all with the traumatic experience, also determines all future experiences. What occurs *after* the psychic trauma has a different weight, a different sense, than it would have had with the objectively identical givens before the psychic trauma. We must therefore free ourselves from the fiction of the isolated individual experience, which lives on as such on the basis of its memory traces or is repressed into the Unconscious. The affective weight of the traumatic experience cannot be derived from the persistence and after-effects of the original event, but rather only from that transformation—which occurs in the first experience—of the experience-world in its decisive and general character.

Even banal events are not devoid of relational connections with the general meanings; they merely call forth no transformation of the system as a whole, and thus appear to the experiencer to be mere repetition, or his attention turns immediately to the infamous species (*infima species*), to the special circumstances of the case. Thus one could compare a traumatic experience to a revolution in which the existing form of state and situation must give way to a new form, sparing no citizen from its impact—in contrast to a banal event, for example, in a government administrative authority, where one official is promoted over another.

The transformation, however, depends not only on the newly emergent meanings but also on the structure of the experience-world into which it is fitted. Only when both factors are in harmony does the transformation occur; only then does the individual experience giving rise to the transformation acquire the character of first-time-ness.

Thus a shocking experience cannot occur in any arbitrary moment in the course of a life; whether or not it comes to pass will be decided not only by external circumstances but also by the condition of the world of the experiencer. And this in turn is determined and fixed already by what is past. Even if a man presents himself as prepared for all shocks, he cannot induce them voluntarily. The law of individual historical development binds him; it no longer allows a repetition of metamorphoses that have once occurred. The metamorphoses occurring in such historically decisive moments expand and mark out the limits of experiencing, and thus outline the contour of the historical figure of the experiencer. The metamorphoses are not only not repeatable they are also irrevocable; they cannot be annulled. Only in fortunate cases may they occasionally be included later in an overlapping expansion of the mental domain and thereby be effectively annulled.

In the contemporary psychological literature, and particularly in the psychotherapeutic literature, one runs across a widespread aversion to acknowledging the irrevocability of historical developments in the course of the individual's life.[15] All decisions appear to the radical psychotherapist to be revocable and reversable. Thus they also assume the character of the temporary and nonbinding. The radical therapist secures his position by an appeal to Freud's drive-theory. The satisfaction of drives is supposed to be discoverable behind all experiences as their open or concealed goal. Drive-satisfaction, however, is the search for a merely objective state, which, resting in itself, removes the historical context and is without historical sequence. The drives have their own immutable goal. As Freud believes, they are conservative and hence actually ahistorical. Their mixtures and alloys are external History[16] recorded by an observer, not inner historical becoming.[17] They are involved in alterations without altering themselves; they lack their own becoming. They are like letters put together by the typesetter into different words: the letters remain the same, untouched by all the word combinations produced from them. The (secret) presupposition of the drive-psychology is that the human being is an ahistorical being and comes under the power of history only reluctantly. What has driven man to this—to construct a different reality than do the animals (who under the influence of their environment never come to the point of history)? The drive-psychology gives no information concerning this question. Reality appears on the scene in psychoanalysis like a *deus ex machina*. Yet man does not merely find or happen upon this actuality that compels men to an only grudgingly borne renunciation of immediate drive satisfaction. Rather this reality, including precisely those parts that harbor the demand for renunciation, is man's own personal creation. To what mental forces does this creation owe its

15. Rejuvenation-therapy transfers this history-blind and life-blind attitude to biological development.

16. Straus uses *Historie*, which I will translate as History with capitalization to designate an external process of one event following another without any meaningful succession, in contrast to *Geschichte*, or history, as a meaningful unfolding of events that stand in an internal and meaningful relation with one another. (Translator's note.)

17. Cf. L. Binswanger, *Lebensfunktion und innere Lebensgeschichte (Life-function and inner life history)*, *Mschr. Psychiatr.* 68 (1928).

origin? To this day psychoanalysis owes us the answer to this question.

In the paper appearing in 1914, "The History of the Psychoanalytic Movement,"[18] Freud mentioned that in a discussion Adler had said: "If you ask where repression comes from you receive the answer, from civilization. If, however, you then ask where civilization comes from you are told, from repression. You can see, therefore, that it is nothing but playing with words." Freud considered this a hair-splitting argument. "What it means is simply that civilization is based upon repressions effected by former generations, and that each fresh generation is required to maintain this civilization by effectuating the same repressions." At that time Freud had not yet recognized that his explanation only pushed back the problem and failed to resolve it. In later works, even quite recently, he has repeatedly attempted to fill this gap through historical constructions. But his efforts have failed to produce success. His historical derivations all consistently and covertly already presuppose the same thing that is supposed to result genetically for the first time from a process of development. The question considering the origins of civilization and the constitution of man as an historical being has thus remained open, and in a naturalistic system must remain open. It is characteristic of the onesideness, but also of the purity and rigor of Freudian thought, that in spite of all of the repeated efforts at this point, Freud has not come any further. Because there can be *no going further* in the direction that Freud adopted and followed. The psychology of drives has no point of approach for a solution of this problem, because the drives are in themselves a-historical. The transition to history characterizes the limit drawn by the psychology of drives. If civilization operates as a regulation of tendencies toward aggression, and if the repressions and limitations on obtaining pleasure proceed from reality, then the forces regulating the life of drives through the mediation of civilization must, along with these drives, belong to the original equipment of the human being; they cannot themselves be derived from the drives. The problem of the constitution of man as an historical being cannot be treated at this point; it will occupy us fully later.[19] Here, where we attend to the phenomena in which first-time-ness shows itself to be meaningful, we must presuppose this constitution as given. The phenomena themselves must compel us to attempt the analysis of that problem.

18. S. Freud, *Collected Papers*, Vol. I, p. 346.
19. Cf. especially Chapter VII.

The idea of a restoration to the *status quo ante* is, as we have seen, based on the antihistorical drive concept; it elevates the nonbindingness and inauthenticity of all action and experiencing to a basic principle—thus negating the meaning of the historical process—and wishes to make an unhistorical being out of the human being. Psychotherapy converges here with widespread, ever triumphant and pervasive tendencies of our day. They are perhaps nowhere more easily comprehensible than in the altered relations of the sexes to one another. In behavior commonly praised as objectivity and veracity, we see manifested the attitude of an individual who does not experience himself as the creator of his own historical figure (Gestalt), but rather as a creature with diminished responsibility toward himself, a creature who suffers and lives through his condition and situations as external forces and internal pressures. Thus when the Catholic Church explains that marriage is indissoluble, by the irrevocability it emphasizes the worth and the meaning of the uniquely occurring decision. It holds firmly to the historical modality, to one-time-ness, and thereby forces the individual to take himself seriously and to fit his actions into a process of self-structuration and self-actualization. Private good fortune could be a result only of fortuitous conditions. Private good fortune—this somewhat home-baked ideal—is held in low regard alongside the general objective form that first makes it possible for the individual to discover his own Gestalt. Casuistry resolves nothing. The good fortune or misfortune of the individual do not infringe on the value of the formations of the objective mind[20] because the good fortune of prosperity, of comfort, or of the satisfaction of vital needs, whether allowed or forbidden, simply does not extend into that sphere.

There remains an essential difference as to whether some, many, or even all individuals offend against (ethical) requirements or whether the idea itself is given up. The altered valuation of virginity, the facilitation of divorce—which has progressed the farthest precisely in the puritan countries—and the marriage legislation of Soviet Russia[21]

20. *Objektiver Geist*: objective mind. In Hegel's philosophy, the absolute mind or spirit actualizes itself in history in the form of the objective mind, that is, law, morality, customs, and institutions. In Straus's anthropological psychology objective mind can be understood as the collective "mind" of culture, which is objectified in the works, institutions, and objective products of the culture. (Translator's note.)

21. In the early years of the Soviet state, marriage was regarded according to the view of Marx's 1844 manuscripts, that is, as an instance of exclusive private ownership of a woman, paralleling the capitalist's exclusive ownership of material property. Thus legislation aimed at loosening the marriage bonds. This legislation included provisions for divorce through the mail at either party's instigation. (Translator's note.)

all have one thing in common: they relieve the individual of the requirement to take himself seriously, and they seek to protect him from experiencing shocks and to banalize his experience—forcing it into the domain of states, moods, and humors. The earlier mood can of course claim no priority over the later. We call a man moody when his decisions are determined essentially by his changing vital states, and in fact especially when these decisions concern that which is objective and historical. Measured according to their validity and currency, the vital states are incidental. The Gestalt-less and anonymous human being is the carrier of this world-view. Unless all of the symptoms deceive us, this process of banalization signifies the transition, notwithstanding the political systems prevailing in individual countries— from the bourgeousie to the proletarian age. We may illustrate how deeply this process has reached by statements from certain investigators who would no doubt link themselves to an entirely different position but whose thinking has nevertheless been decisively influenced by this historical transformation.

Prinzhorn writes in his *Psychotherapy*:

> . . . proceeding from this most extreme biological standpoint or sub species of the vital meaning . . . one knows at the same time why . . . a deep, paradoxical and enduring tension must prevail between the breakthrough of the urge to love and the scattered efforts at sustaining society and inventing reasons (not accessible to me) for the merit and dignity of chastity.

Here, as at many other points in his book, Prinzhorn points to animals whose world-certainty and biological security he sets forth as a model for man. But this reference is entirely misplaced. It does not develop the teachings of Klages further; rather it coarsens and distorts them.[22] The animal lives through states of passion that recur just as hunger and fatigue recur, and find their instinctive satisfaction just as hunger does in the reception of nourishment and as fatigue does in sleep. The first passion and mating for the animal is nothing more and nothing different than the following ones, just as sleep and nourishment this day mean nothing more than those the day before. The human being, however, does not merely live through states of sexual excitement; rather, in satisfying them man experiences life—his own life and the life of the

22. L. Klages, *Prinzipien der Charakterologie* (*Principles of Characterology*), 3rd. ed. (Leipzig: 1921).

other. As experience, the submission and unification acquire an historical meaning in the life of the individual, and through this historical meaning, because the shock is bound to first-time-ness, the first incidence receives and preserves an absolute priority above the later incidences. The change in the valuation of virginity, therefore, absolutely cannot be understood as a changed, unprejudiced, and natural attitude toward sexuality; it must be understood rather as a banalization of self-structuration or as a far-reaching renunciation of this. In this area of sexuality, the shock results from the experiencing of objective first-time-ness; there is a compelling bond here between *event* and *experience*.[23] What has once occurred can no longer be effaced; the past does not return; there can be no restoration of the *status quo ante*.

Even the successes of psychotherapy cannot rest on such a restoration. Contending parties can certainly close an agreement in which the status quo is proclaimed, as in a peace treaty, but falling-back to old forms and formulae cannot take away what has happened in between. After a bitter dispute one can extend one's hand in reconciliation; things nevertheless will never again be the way they were before the dispute. The Bourbons, following their return to power in the years 1814 and 1815, could endeavor as much as they wished to restore a tie to the old times, but the transformations that had occurred could not be annulled, and the events of the past years were not to be erased from the memory of men, although, as Stendahl tells, it was forbidden in many salons of this restoration time even to mention the name Napolean.

The shock thus depends on the first-time-ness, and this is based on the uniqueness, directedness, and nonrepeatability of the historical process. In the search for impressive phenomena that can elucidate this relationship, the psychopathologist—in contrast to the historian and biographer—tends to choose cases in which the irrevocability of the historical occurrence gives rise to characteristic disturbances of behavior. As a final example of the significance of the historical modality, we will single out and briefly analyze one more phenomenon from this group—avarice.

The miser is the man with the cash box, as Moliere has portrayed him. He buries his treasure in order to save it from the clutches of enemies and friends. But at the same time he deprives himself of it. He

---

23. Concerning the question of compulsion toward deriving a certain meaning, compare the later discussions in Chapter V.

spends an entire life in acquiring, securing, and watching over his possessions, without ever enjoying its pleasures himself. On the contrary, he sacrifices and wastes his own days merely to prevent the use of his fortune whose sole worth is realized only in its utilization. From time to time we hear that somewhere a man has been discovered in the most impoverished surroundings, half starved and decayed, and that after his death the astonishing discovery is made that he possessed great riches. Such pieces of news, whether true or false, reveal the true nature of the miser. It is not enough to say that the miser wants to give away nothing; wanting-to-give-away-nothing is the mark of the small-hearted. Certainly the miser wants to give away nothing; but this is only a result of his opposition to any and every use of his fortune; he even resists when his money is to be used for himself. Wanting-to-give-nothing-away could mean that the miser refuses to renounce his claim to any possession, for the benefit of others, without compensation. But he even favors himself with nothing; "he sits on his money"; he will not even give out his money when he could exchange it for an appropriate equivalent. If Moliere's miser has possessed an empty cash box, he would not objectively have been any the worse. He exerted all of his strength to defend himself against utilizing those possibilities presented to him by his fortune. With tenacious obstinancy, he held firmly to the bare *possibility*, and resisted every actualization.

The battle against actualization—the perseverance in a state of possibility—is the decisive factor in the nature of the miser that binds him to the money. All material goods are consumed in use; in fact they almost all gradually perish even if they remain withdrawn from use. Only money is not exposed to this process of arising and perishing. It has duration, it retains its value.[24] Money is objectified possibility. Its durability, its potential power, and its anonymity bind the miser to the money, and only this bond renders his behavior entirely intelligible. What is decisive besides the durability, is the potential power of the money. Its meaning for the miser can be made intelligible in a simple example.

If someone possesses 1,000 dollars, as long as he does not give them out, he can buy this and that and a third thing, in fact almost any number of individual objects that he chooses. But this is true only as long as he

---

24. The exceptional conditions of times of inflation do not invalidate this interpretation of minted money.

does not actually put this into practice. So often those who unforsee-abiy acquire a great sum through inheritance, gambling, or a lottery are led astray—by the possibility connected with money—into foolish and rash squandering. When the riches suddenly become their lot they experience an almost immeasurable expansion of their "capacity" (or fortune). They overlook, however, that the abundance of possibilities stands at their disposal only so long as they have not yet given anything out. In the same moment, however, that one buys something for this 1,000 dollars, the possibilities shrink almost into nothing. There is a transition from the boundless quantity of purchasable things into nothing more than one or two objects now in his possession. The fantastic possibilities are replaced by narrowly limited and modest reality. This process of shrinkage, which characterizes every transition to material reality, is certainly nowhere more tangible than with money and its utilization. Thus many who go forth full of delight to buy themselves something or other almost as a matter of course come home feeling dejected simply because in the moment of decision nothing remains from the abundance of enticing things open to them before the choice, except the one hat or pair of boots that they bought for themselves. On the way home they now find all of the objects they were forced to renounce doubly alluring. They are sad: the renunciation and the loss of possibilities depresses them more than the acquisition gladdens them.

The purchase of hats and boots is a trivial example, but it shows—in a crudeness desirable here—the necessity of renunciation and of limitation in the transition from the possible to the actual, a necessity that does not become equally clear in all modes of behavior.

It is especially easy, because of its relationship to money, to comprehend greed as a form of resistance against actualization, that is, a form of the tendency to abide in the state of possibility. This resistance is a universal one with the miser, but it operates in a particularly characteristic manner in his bond to money, whereby the acquisition and securing of a fortune lends a positive meaning to his behavior, at least to all appearances. Resistance against actualization is the basic tendency of the behavior not only of the miser but also of the compulsive patient. This is true so long as we speak of the original compulsion states and not of the accompanying depression. Resistance against actualization, against historical becoming, unifies the diverse manifestations of the syndrome described by psychoanalysis under the designation of the anal-erotic character, but at the same time refers these manifestations to another place in the structure of the personality. The psychoanalytic discussion of the anal-erotic character is based on the symbolic equivalence of money and excrement. This equation can be interpreted in two different manners: first, that excrement is valued as possession; or second, that money is regarded as dung. The psychoanalyst holds the first interpretation to be the correct one. This, however, overlooks the fundamental distinction between the two members of the equation. There is a vast difference between defending an already held possession with a characteristic value of its own against alienation and theft, and defending a possession that is only a means to a possession with value of its own. In the one case the defense directs itself against the perishing; in the other, against the becoming. While holding fast to property that is already possessed can still belong within the domain of the instinctual—even the animal defends his booty— holding fast to the means and congealing in mere possibility have their place only in an existence that is experienced as specifically temporal and as historical becoming.[25]

Thus the resistance against self-actualization that is exhibited in greed points to the organization of this historical process, to its being-directed, its first-time-ness, and its irrevocability. The irrevocability discloses the characteristic nature of the historical becoming in a backward glance at the past, as the first-time-ness does in a forward glance at the future.

The demonstration of first-time-ness as an essential moment in a shocking experience places new tasks before us. In the attempt to distinguish and to define classes of experiences and to inquire after the conditions for their actualization, one constantly runs into unnoticed, unsettled, or unresolved problems of general psychology. The deep-seated transformations that psychology has undergone since the beginning of the century have compelled us finally to abandon that departure point for all psychological investigations, which was characteristic and self-evident for the old research orientation. The delineation, description, and analytical penetration of definite classes of phenomenally simple experiences—or of experiences theoretically conceived as

---

25. Many features of the anal-erotic character such as pedantry, the love of order, and spite are also not infrequently found in the spendthrift. The bond to the *means* is common to the spendthrift, the collector, and the miser; through this bond the miser is distinguished from the "man without means," with whom he often shares the burdens of poverty.

simple—has ceased to be the first, and should now become the last task of psychology.

If we nevertheless attempt to analyze individual classes of experiences, then just when we believe ourselves to be nearing the goal it will all at once move again into the remote distance. So it has gone for us now. With the aid of the concepts of representative function and of first-time-ness we hoped to be able to master theoretically the peculiar nature of a shocking experience, yet now we discover one more kernel of unresolved problems within this first-time-ness itself. We cannot undertake an attempt at their resolution here because this would force us to confront and discuss the basic problems and basic principles of general psychology. Here we can only mention the decisive question posed by the reference to first-time-ness.

The concept of first-time-ness signifies that the historical process, in which first-time-ness operates to organize experiences, must be *directed alteration*. Only if the individual experiences form members of a series and if they are governed in their content and in their existence by the principle of the series, can first-time-ness take on the significance we have here ascribed to it. The concept of series is intended here in the strict sense and therefore should be sharply distinguished from the psychological term "series of images." There is a sharp contrast between the *chain* of associations and the series. At any given time only two adjoining members in a chain of associations can be considered to be connected with one another, while in a serie all members are together subject to the lawfulness of the series that rules their collective course uniformly.[26] The members of a series thus stand in a pervasive relationship to one another.

It may appear unobjectionable at first to employ the ordinal numbers to characterize the positions in a psychic series. However, one should not deceive oneself about the lack of absolute harmony between the principle of a psychic series and that of a series of ordinal numbers. While in this ordinal series the step from the first to the second member of the series is the same as that from the third to the fourth, or from the ninety-ninth to the one hundredth, the members of a psychic series are different in rank and meaning, according to their position within the

26. I pass over the productive critique that H. Liepmann has made of the concept of a series of images since, in our meaning of the term, his work on the flight of ideas has not yet posed the problem of the psychic series.

series itself. It is possible to refer to increasing quantities in the ordinal number series; this is not possible in the psychic series. We must heed this distinction between the two series or we will once again unwittingly introduce, along with the ordinal numbers, a quantitative mode of thinking into psychology. The employment of the ordinal numbers facilitates understanding to be sure, but we may not overlook that in using them we still have not resolved the problem as to the principle governing a psychic series.

Can psychology resolve this problem? This question should be answered in the affirmative. To the extent that psychology is an "understanding psychology,"[27] it should resolve the problem; this problem belongs to its task-area. But if we inspect psychology's contributions to date toward the solution of this task, we must confess that there are few cases indeed in which psychology has even heeded the problem.

Such expressions as the "stream of experiencing" certainly allude to the directed alteration, but even in William James's description of the stream of experiencing the question concerning the psychic series is neither precisely posed nor in any way answered.

Binswanger has already shown what phenomenology can accomplish toward resolving this question.[28] In the life history

> [he says], the inner essence of the individual, his mental person, unfolds and forms itself, and inversely, we first learn to know the mental person from the inner life history and only from it. This process of learning to know the person is the historical-hermeneutic exegesis or interpretation.

This psychological exegesis of the individual person has as its basis an entirely fixed realm "of mental being," an entirely fixed unity "of inwardly self-required moments of meaning," or an entirely fixed unity "of inner motivation-formation." Binswanger adds, "This special province of mental being is the region of the pure relationships-of-essence of psychological motivation, and thus of eidetic psychology, whose exploration is the task of the pure phenomenological hermeneutics or of the pure investigation of essences." As Binswanger has

---

27. Wilhelm Dilthey (1833–1911) introduced the idea of a human scientific "understanding psychology" in contrast to a natural scientific "explaining psychology." Understanding (*Verstehen*) aims at describing and comprehending the total and unitary structure of man in its individuality, while explanation (*Erklärung*) aims at giving categorical reasons or causes to account for an event's occurrence. (Translator's note.)

28. Binswanger, *Lebensfunktion und innere Lebensgeschichte*, p. 69.

emphasized even more clearly elsewhere, this preoccupation with inner life history involves reflection on a *mental* connection and pursuit of a *mental* "being." Accordingly the life historical *becoming* would not itself be the object of an ontological investigation.

Thus Binswanger does not even treat the question concerning the principle of a psychic series as an appropriate problem. The problem is simply not posed and probably cannot be posed at all by phenomenology.[29]

Only by abandoning the "psychology of elements" has it been possible to make the comprehension of mental being as an *individual historical continuum* into an object of psychological investigation. The development toward this goal, however, has been delayed by the effort of Gestalt psychology to substitute, in the place of the old *sensation-atomism*, a *Gestalt-atomism*. Through this Gestalt atomism, psychology has recently been in danger of falling under the power of an extreme physicalism. The natural philosophic attitude of the Gestalt psychologists, and their tendency to dissolve the distinction between the "inside" and the "outside", brings the psychic series into an unequivocal dependence on the stimulus and on the process of biological events. Certainly no one will dispute that the psychic series stands in correlation with a series of events and with biological development. Even to survey this correlation completely, however, it is necessary to call into question and to work out the inner lawfulness of the psychic series. Neither the external extent nor maturation and aging alone determine the before and after of experiences. Experiences are not arbitrarily interchangeable. The inner principle of a psychic series prescribes a certain direction and sequence to them, independent of the succession of events and of any alterations of the life-functions.

For the historiological mode of thinking "the New" becomes a problem, indeed a basic problem and departure point for memory research in general. We hold as an axiom that only the New can be incorporated as the property of memory. The performance of memory

29. Although Straus drew heavily from Husserl and other phenomenological sources, he did not identify himself closely with phenomenology as a school and movement until after his emigration. Earlier in his career he referred to his approach most often as *historiological*—as in the present monograph with its emphasis on time, the process of becoming, and the historical nature of the human being. He also regarded himself as a spokesman for the *anthropological* psychiatry espoused in *Der Nervenarzt*. In this country he did not adhere so strictly to these distinctions and accepted such loose appellations as "existentialist" and "phenomenologist." (Translator's note.)

is thus not confined to apprehension, retention, and reproduction;[30] rather it begins with the formation of the material—the stuff—which will be admitted to apprehension. Memory does not encounter a definite something that it would be its task to notice; rather it forms that something through a process of organization from the stimulus-givens and according to its own systemic conditions. The dispositions or the engrams are not formed in a simple dependence on the stimulus, they also do not originate through an amalgamation of the actual sensation-material with earlier images. Only the New can incite the formation of dispositions and engrams. The New belongs to the content of experience. Its character as a relation shows us that apprehension, retention, and reproduction do not take up just any arbitrary, static particulars; apprehension is instead already systematically conditioned. Something can be New only in relation to something else, different from itself, and only in relation to an historical continuum. Thus the New depends on the one hand on the system of objects, and on the other hand on the self-completing system of the individual, historical person. This demonstration of an organization of the material of memory, systemically conditioned already through "apprehension," for the first time makes it possible to explain the systematic character of memory, which obtrudes itself upon us clinically, especially in systematic deficits such as the Korsakoff syndrome[31] or retrograde amnesias.

If our presentation of the historical modality is not to remain incomplete and inconclusive, we cannot omit a reference to the relationships existing among first-time-ness, the psychic series, and the concept of the New. We were forced to confine ourselves here to a brief discussion and could not achieve a complete overview on the totality of questions posed here. Only in a forthcoming work will it be possible to investigate and to present in appropriate detail the range of problems concealed in the concept of the New.

30. Apprehension, retention, and reproduction: the three phases of memory function in the classical memory theory. (Translator's note.)

31. Korsakoff syndrome: an organic brain disorder commonly found in alcoholics and characterized by amnesia, particularly for recent events, secondary confabulation (filling in the memory gaps), deficits in attention, disorientation in time and place, and multiple twitches and pains secondary to peripheral nerve degeneration. (Translator's note.)

3 / Toward a Critique of Genetic Theories

Earlier experience governs the phenomenal form of later experience, childhood esperiences prefigure the plan for the later life-formations; one image can appear vicariously for another; and an affect can be displaced from one image to another.[32] All of this is nevertheless possible only because the original experience has already been a *representative* one. Only through mediation of the general meaning, actualized for the first time in the earlier experience, is the later experience bound to the earlier. The genetic mode of thinking that conceives the first experience as a single experience and the later as its mere reproduction is not even in a position to render comprehensible how the later experience could take on the character of reproduction.

Our concept thus stands in sharp contrast to those theories in psychology that wish to derive and to explain the class of an experience from the constellation of circumstances in which this kind of experience appeared for the first time. Our critique is not directed against concepts of development such as that of Werner, in his *Introduction to Developmental Psychology*.[33] When Werner characterizes differentiation and centralization as the essence of mental genesis he merely expresses the logical-systematic principle of development. In general, wherever mental and organic formations can be ordered into a series of increasing differentiation, we assume there is development. In doing so we project the series, which is systematically ordered according to an ontological principle onto the level of historical becoming. The more

32. I have chosen the expression "affect displacement" here because it is a term already brought into usage by psychoanalysis. Actually I consider the ideas expressed by this expression to be false and misleading.
33. H. Werner, *Einführung in die Entwicklungspsychologie* (*Introduction to Developmental Psychology*) (Leipzig: 1926).

highly differentiated formation is thereafter interpreted as later temporally. In considering phylogenesis, we are not guided by any actual experience or observation because the developmental stages were traversed one at a time. Instead we construct the development by translating the logical system of differentiation into a real one of self-differentiation. This still tells us nothing at all about the forces causing the development. In the genetic theories that we combat, on the other hand, development is conceived as an explanatory principle of differentiation, and development becomes a process of origination, which in psychology has a mixture of causal and intelligible connections as its presupposition and consequence. Relationships concerning the structure, and therefore the suchness (*Sosein*) of the experiences, are interpreted as sufficient ground for their existence (*Dasein*). Inversely, experiences that have been factual or presumed conditions for the appearance of other experiences are interpreted as constituent parts of the later experiences.

According to Freud, anxiety originated through the birth trauma, and affects *are* the *reproductions* of old, vital situations. That the original situation is meant entirely in the sensualistic sense hardly needs further emphasis after what has already been said. The first experience is conceived as a one-time natural event of limited duration. It consists of a succession of sensations and certain pleasure-tones and unpleasure-tones of the sensations; it is originally devoid of sense, meaningless—a bare objectivity and neutrality. What appear to be sense, meaning, and systematic relationship are the result of the real after-effects exercized, with the help of the memory traces, by the experience. Later these memory residues are again activated by similar external stimuli; they then bestow upon the later experience the imprint of the earlier.

Actually it should suffice as a critique of genetic theories, to refer to the works dedicated to overcoming psychologism in logic, because basically we are dealing with questions similar to those addressed there. The great respect that genetic psychology now enjoys, however, shows how little one district of scientific research takes notice of what happens in a neighboring district. While the renovation is already fully in progress on the ground floor, someone on the second floor is still delighting in the splendid old architecture.

Genetic theories completely overlook the problematic of the single experience, which was pointed out in the preceding section. The single experience is still treated as a relatively simple formation and is still viewed as the self-evident departure point for psychological investigation.

The inadequacies of genetic theories could be disguised this long only through the use of expressions such as repetition, connection, and reproduction, which have entirely different meanings in physiology than they do in psychology; they have been used side by side with continuously alternating meaning, as if the same concepts were being applied each time.

What can reproduction mean in general? If I see the blue of an object today, which I had also seen yesterday or three days ago under exactly the same physical conditions, may I then say that my sensation today reproduces yesterday's or the earlier day's? Would it be legitimate to say that today's sensation is the same as yesterday's? Plainly not. Because the sensation as a natural event has a beginning, end, and duration in empirical time. With its conclusion at a definite point in time it is unrestorable. Only an extremely inexact mode of expression permits one to speak of the repetition of natural events. A conductor who during an orchestra's rehearsal demands at the end of a movement of music that the musicians repeat the last ten phrases, intends a renewed performance of a certain musical form. Thus the notes resounding are not in themselves a repetition; rather, the earlier as well as the later are once-occurring phenomena bound to their temporal position. They first become a repetition through the fact that the experiencing human being—and that means the historically experiencing human being—apprehends the notes as representatives of the one musical theme. The situation is no different for the perceptions in the example introduced above. The later perception repeats the earlier only insofar as both have a certain species "blue" as their object. We may not say therefore that the same sensation returns, rather only that a similar one is given again. However, we have recognized that the two sensations, the earlier and the later, are different, and are comparable only in a certain respect. Every reproduction rests on a comparison and requires a *tertium comparationis*, which connects the two members of the comparison. The two perceptions of blue surfaces, separated from one another by a temporal interval, are thus comparable only to the extent that both times the same species blue appears to the historically experiencing individual. Notice, however, that it is not the individual object but rather the species—represented by it and attaining a manifest givenness in it—that forms the *tertium comparationis* and thereby the condition for reproduction. Thus there is reproduction only for an historically experiencing being, and only to the extent that the individual experience is representative for a species or general meaning. Were the first experience not already a representative one rather than only a singular one, it would absolutely not be reproduceable. With respect to the representative function, the earlier and later events cannot be distinguished in any way.

Perhaps it is necessary to show the impossibility of genetic theory more precisely through specific examples. I choose as the first example Freud's attempt to derive anxiety from the birth trauma. Anxiety, Freud believes, originates as a reaction to a condition of danger.[34] Birth, objectively considered, is a danger for mother and child. The birth process calls forth in the child a series of sensations and motoric innervations, one after the other.[35] "Anxiety," it is said in a first approximation, is thus "a special state of unpleasure with acts of discharge alng particular channels." Sensations and innervations— which can occur and vary independently of one another and are called forth by different kinds of stimuli—are unified through birth into a fortuitous, physiological ensemble. Freud's entire aim at this point is to demonstrate that the experience of birth is constructed from sensations and innervations, which were first understood entirely as bare, neutral objectivities. Once this construction is accomplished it should be possible that in later experiences of danger, inversely, anxiety will be reproduced in its specific individuality, characterized by unpleasure-affect and by motoric discharge. Thus we see that even genetic theories cannot manage without a *tertium comparationis* drawn from the meaning-sphere of the experiences. Freud's presentation is genetic only to the extent that he describes anxiety as coalescing from isolated, individual sensations under the influence of a particular physiological event, but at the same time he also provides this ensemble of sensations with a definite meaning-character that first makes possible the reproduction.

Freud's leap from the domain of the sensation into the sphere of meanings was successful or, to be more accurate, *appeared* to be

34. S. Freud, "Inhibitions, Symptoms, and Anxiety," *Standard Edition of the Complete Psychological Works of Sigmund Freud* vol. 20 (London: Hogarth Press, 1926).
35. I pass over all of the empirical objections against Freud's hypothesis and, to begin with, entirely follow his description since a methodological discussion is possible only in this way.

successful, only because he used expressions that have an objective sense, for example, the increasing of the "discharge requiring stimulus magnitude," in such a way as though he were engaged in a physiological description of contents of sensation and their alterations. That alone made possible for him a continuous, unnoticed, and in his case necessary alternation between the two modes of thinking—the objective and the content oriented.

The disunity in the effort to derive anxiety genetically is already inherent in the fact that on the one hand the experience is supposed to be constructed directly from the sensations, and on the other hand the experience is supposed to bind the sensations and innervations one to another. Even Freud believes that the purely physiological combination cannot suffice:

> . . . we are tempted to assume the presence of an historical factor which binds the sensations of anxiety and its innervations firmly together; in other words, that an anxiety state is the reproduction of some experience which contains the necessary conditions for such an increase of excitation and a discharge along particular channels, and that from this circumstance the unpleasure of anxiety receives its specific character. In man, birth is a prototypic experience of this kind, and one is therefore inclined to regard anxiety states as a reproduction of the trauma of birth.[36]

One could hardly claim at this point that the investigation, which is so decisive for the entire theory, has been carried through in great depth. The mode of expression is figurative and indefinite, and the most important questions are not even posed, let alone answered. Does the historical factor of which Freud speaks operate only once and then drop out of further experiencing, or is it also a factor contained in the reproductions? Is the historical incident formed by the biological processes—that is, the succession of pains, the squeezing through the birth canal, the break-through of the head, and so forth—or is the birth meant to be a conceptual compilation of the biological processes, the birth as *one* incident, or, finally, the experiencing of birth as a separation from the mother? How does the historical factor bind the isolated sensations and innervations to each other? How are the elementary constituent parts altered by this binding? Is this merely a matter of factually and temporally appearing together or appearing one after the

---

36. Freud, "Inhibitions, Symptoms, and Anxiety," Ch. VIII.

other, or do the sensations and innervations also acquire a new intrinsic meaning through the binding?

For the natural attitude as well as for the natural philosophical mode of thinking, the individual processes taking place with birth and extending over a long temporal interval, are parts of a unitary event. Only as such can it be an historical factor—that is, more correctly an historical formation—and only as such could a binding power be attributed to it. The individual processes elapsing one after the other, on the other hand, cannot draw together and bind because they themselves are and remain isolated. One could also dissolve a military battle into a thousand individual activities of the participants. One could think of using a complicated apparatus to register all of the movements carried out by individual soldiers, all of the words uttered, and all of the physical events: shots, explosions, gassings, and so forth. The sense of the historical incident, "battle," however, could not be distilled from any of these details, nor even from their sum. The sense abides in none of the individual processes; indeed it exists only as an order overlapping all particulars. The individual frames of a film strip know little of each other; the succession of frames in itself hardly contains the event seized in the initial act of photography, and the individual phase, or the individual position during birth, does not in itself contain the sense of the birth. Thus, if the historical formation is actually supposed to be able to bind, then either the objective sense or the experienced sense of birth must already be given, because it is the sense that first draws those particulars into a unity, which in turn can unite yet other scattered particulars. Freud seeks to circumvent this difficulty by taking something that has been expressed by the concept, "the increasing of the discharge-requiring stimulus magnitude", that is, the sense of the event, and projecting that into the sensational givens. The genetic theories are able to sustain their position only because they already insert into the sensations a sense that bare sensations simply cannot have.

The Freudian train of thought is clear: birth, objectively considered, is a danger. The newborn knows nothing of that. Even Freud does not expect the newborn to actually experience birth as a separation from the mother. Rather, a wonderful, prestabilized harmony ordains that the individual physiological processes of birth (which in themselves contain nothing of danger and are nothing more than individual physiological processes), by their impression on the organism of the child, will call forth sensations in the child. This harmony also ordains that these sensations will then be ordered in a series of increasing stimulus magnitudes and thereby mediate to the child the experience of danger. If birth is objectively a danger to the extent that it can miscarry and can destroy the life of the mother, of the child, or of both, then in the psychoanalytic sense the experience of danger arises precisely with success in birth. But this is only a relatively unimportant displacement of the accent of meaning. What is decisively important is that Freud overlooks the fact that the expression "the increasing of the discharge-requiring stimulus magnitude" articulates the ordering principle of the series, which orders the successively occurring and subsiding sensations. However, this order itself is no longer a property of the sensations as such. This is a decisive distinction: whether the *increasing* of the sensations is experienced, or whether it is the increasing *sensations* in their successive appearance that are experienced. The two expressions sound so similar; nevertheless their meanings are separated by an unbridgeable chasm. If he wishes to remain true to his basic principles, Freud may speak only of increasing sensations because the increase and the "discharge-requiring" quality are not mere sensational givens.

Let us assume that the sensations occurring in the child during the birth can be ordered into a progression of growing intensities. I_1, I_2, I_3. Sensation I_1 has already faded away by the time sensation I_2 appears and the same holds for I_2 with the appearance of I_3 and so forth. If, following the sense of the genetic theory, we take the individual sensation strictly as individual, as such it refers neither to the preceding nor to the following sensation.

In general sensations can be experienced as increasing only if they are comprehended as the members of a temporally ordered and directed series. The first words of a sentence have faded away when the last is spoken, but the sentence is nonetheless experienced not as a succession of words but instead as a sense unfolding in and with the individual words. In the same way the individual sensations, called forth one after another by the processes in the maternal organism, do not bring forth the ordering principle of the series out of themselves.

According to Freud, however, the sensations are experienced not only as temporally ordered but also as definitely directed. Anxiety arises not because in the end a sensation of a particular, heightened intensity occurs; it arises because the series has the *direction* of increasing and therefore of danger. Yet increasing in itself does not require discharge; neither does it imply the character of danger. If a dark space were to gradually become ever more illuminated, increasing quantities of stimuli will break in upon the organism; but this increasing is perceived pleasurably by everyone who is awake and wants to see. I do not want to go into any more detail here about one of the fundamental Freudian hypotheses, which maintains that the human organism is directed toward minimizing tensions, and that an increasing of tensions results in unpleasure, and tension reduction results in pleasure. In this example it is important to me only to point out that "discharge requiring" and anxiety cannot yet result from the increasing alone; it can result only from the threatening *meaning* that the increasing has in this particular case. The sensations must therefore be organized according to a temporal principle. They must be directed in a series of growing intensity, and added to this there must still be a specific quality; only the temporal order, the direction, and their specific quality can together lay a foundation for the experience of being threatened.

The individual sensation as such does not require discharge; it does so only by virtue of the position it takes in a series ordered by a specific principle. One need only recall at this point that the physically identical temperatures of the surrounding medium are perceived as cold if one comes from a warmer environment and, inversely, as warm if one comes from a colder environment. One need only consider the series inversely, so that the ascending line becomes a declining one, and then every member of the series receives the inverse sign. Each sensation thus will operate entirely differently according to the character of the series in which it stands even though, considered individually, it possesses the same quality and intensity. If a pain that we have suffered—for example, by pinching a finger—begins to fade, the fading away is experienced as pleasurable or—I would rather say—as a blessing, even though the sensations now occurring are in themselves still sensations of pain. It still hurts, but the series of sensations produced during the fading away—a series presenting a return to the condition of indifference—becomes a positive and a "bitter-sweet" experience. Once again we find confirmation here for the opposition raised by so many researchers in recent years against the assumption of isolated sensations.

Sensations thus certainly do not bring forth the series out of themselves; they are ordered into a series since every sensation appears from the onset only as a member of a relation, that is, as an alteration of the sensory sphere. Even when no continuously increasing or decreasing series is present, the sensation is nevertheless a member of a relation. The existence of a relation is especially easy to perceive in the increasing and the decreasing, because both continuous increasing and continuous decreasing require a relation of at least more than two menders. The objection has already been legitimately made against Fechner that the sensation of a transition is not a transition-sensation. Correctly understood, the Weber-Fechner law signifies precisely that every sensation is experienced as an alteration of the sensory sphere.[37] It can therefore also only take place *by a leap*, that is, discontinuously. It is of fundamental importance for the insight we are seeking here that the sensation is experienced as an alteration of the body-environment relationship or of the body-I relationship. Irrespective of which modality or quality is involved, the sensations have the character of an event; they are directed and appear in relation to a before and an after. It follows from this that the increasing or decreasing series is made up of *discrete* members and that they must be changeable, or inconstant. If the series were constant, with suitably chosen experimental conditions the increasing would not be noticed at all. Thus Preyer[38] could show, in experiments on frogs, that with sufficiently slow augmentation of the objective stimulus strengths, the experimental animal failed to react altogether. The chemical and thermal stimuli could be increased to the point of death in the experimental animal, and the mechanical stimulus to the point of crushing the limbs, without producing any defense reaction. Froebes seeks to explain this behavior by saying that only a faster arousal of the pain would be a stimulus for the instinctive action. He fails to see that in this dependence on time the sensations have become integrated parts of a directed event process.

The findings of Hall and Motora, as well as those of Stern, which all deviate apparently from Preyer's results, can also be fit into this interpretation without doing violence to them. Hall and Motora found

37. The Weber-Fechner law, resulting from Weber's experiments in psychophysics and from Fechner's efforts at refining and quantifying Weber's findings, states that when one distinguishes between two stimuli, it is not the absolute difference between the stimuli that is perceived but the ratio of the difference to the intensity of the first stimulus presented. The ratio of this just noticeable difference (jnd) to the first stimulus' intensity (I) is a constant (k) for any given type of stimulus: jnd/I = k. This law is not valid at the high and low extremes of the stimulus range. (Translator's note.)

38. Cited according to Froebes, *Lehrbuch der experimentellen Psychologie*, (*Handbook of Experimental Psychology*) 2nd and 3rd eds., Vol. I (Freiburg: 1923), p. 416.

that "with pressure alterations climbing at different rapidities, the slower the alteration the smaller was the threshhold." This is easily explained if one were to consider the event-character inherent in the sensations. The experiments of Hall and Motora can be presented graphically as a step-shaped rise of a curve, in which, according to the longer or shorter duration of a stimulus before its alteration, the individual steps would be rendered as wider or narrower. A pressure of a particular intensity is an alteration with respect to the preceding condition. If one were to quickly strengthen this pressure by just a trifle, the second alteration would be unimportant in comparison to the first. If, however, one were to allow the pressure to act for a longer time, so that a new *niveau* (level) would develop, even a slight augmentation of the stimulus would already appear as an event, that is, as *an alteration of the niveau*. In the immense literature on the Weber-Fechner law, as far as I see it, sensations have always been treated as static and isolated. Their directedness and their appearance in the sphere of sensations as a becoming and as a confrontation between I and world have not been considered, although just in the last few years the measurements of chronaxy[39] have already suggested just such an interpretation. Time thus plays a role neither in the familiar formula of the Weber-Fechner law nor in its interpretation—not even in the psychological explanation of the law given by Wundt, which attributes the diminishment to the comparison of the sensations mediated through apperception. Thus, surprisingly enough, no one has ever even attempted unification of those phenomena treated under the title of "perceptions of change" with the phenomena on which the Weber-Fechner law is based.

The historical structure of sensations has a further result in that every sensation possesses inherently both a gnostic and a pathic moment. In the lower senses, particularly in the sense of smell, the pathic predominates, while in the higher senses, particularly the sense of sight, the pathic recedes ever more behind the gnostic factor. Nevertheless a sudden increasing of the optical stimulus—a sudden transition from dark to bright—allows the pathic moment to again come to the fore even in the optical sector of the sensory sphere. Under circumstances such as a blinding or dazzling, this can occur with such strength that the

---

39. Chronaxy: a term in psychophysics, it is the length of time that an electrical current twice the magnitude of the rheobase is required to excite a nerve fiber or muscle. The rheobase is the minimum current necessary to stimulate a nerve or muscle. (Translator's note.)

gnostic disappears almost entirely behind the pathic factor. We then comport ourselves in relation to visual sensations in the same manner that we otherwise do only in relation to smells. In touching, the impressions of touching and of being-touched can most easily be exchanged with one other. Every touching is also a being-touched. Perhaps nowhere other than in the tactile sense is the revelation more distinct: the sensation is an "eventing," a process in the confrontation of I and environment (Umwelt).

The pathic indifference, which led Stumpf to assume a special class of feeling-sensations,[40] and Foerster to assume an affective sensibility,[41] is only an artifact of the experimental conditions common to all experiments concerning sensory-sensations. The situation of the experiment keeps the confrontation of I and environment in suspension as it were. Nothing in it is altered as long as the experiment lasts. As is so frequently the case, the nothingness of indifference is actually only a nothingness of meaning.

Stumpf believes that a sensation can be entirely or nearly feeling free (indifferent) without approaching its zero point. In a similar way Foerster emphasizes that in the normal state the movement sensations arising in the execution of a gentle, smooth, passive movement are as good as free of every affective component; according to his own perception they are a fully affectless sensation. Both authors erroneously equate *affective indifference with affective nothingness*.

Remodeling Henry Head's distinction between epicritic and protopathic sensibility,[42] Foerster therefore believes it necessary to assume a morphologically and functionally separated system for the pathic moment of sensation. He characterizes this system as the affective system of deep sensibility and opposes it to the perceptual-epicritic system of skin sensibility and deep-sensibility. While the stimuli of the epicritic system furnish sensations with the indications necessary to utilize all the details of the stimuli—their origin, their location, their strength, and their combinations and succession—the sensations of the affective system are characterized by a "latency of arousal for the

---

40. Stumpf, *Uber Gefühlsempfindungen* ("Concerning Sensations of Feeling"), *Z. Ps.* 44 (1907).
41. O. Foerster, *Die Symptomatologie der Schussverletzungen der peripheren Nerven: Handbuch der Neurologie*, (*The Symptomatologie of Gunshot Injuries to the Peripheral Nerves*), Supplementary volume, Part 2 (Berlin: 1929).
42. According to this distinction, *epicritic* sensation provides information to the organism, while *protopathic* sensation is affective. (Translator's note.)

feeling process, great intensity of the same, with a tendency toward defense reactions, toward perseverations outlasting the original stimulus and with periodic increases and decreases, toward periodic return after one or more phases of feelingless intervals, toward a lack of location-signs, and toward diffuse spreading and irradiation."

It seems to me that Foerster overlooks the psychological problem concealed in this distinction and seeks to explain transformations in the structure of body-perception as elementary differences between different stimulus-conducting systems. The specific sense-energy of the affective system of nerve fibers expresses itself in the affective sensations. On the other hand, the sensations mediated by the perceptual-epicritic system are supposed to be entirely lacking in every feeling-tone. Foerster uses the following argument to support this concept: under normal conditions in an intact perceptual-epicritic system, a pressure of a particular strength will produce only a pressure-sensation without any noteworthy accompanying feeling-tone. If one disconnects the epicritic system, however, this same pressure under certain circumstances will evoke a frightful pain. In the first case the epicritic system hinders the arousal of an inadequate feeling-tone; in the other case this inhibition is omitted and the affective system reacts in an unbridled manner and strength. The inhibition is naturally understood here entirely in the physiological sense. It seems to me, on the contrary, that, without prejudice to any findings that the physiological experiment or anatomical investigation could produce in this area in the future, only a psychological explanation within the context of the entire theory of the sensations will do justice to the facts of the matter.

Anyone who has once gone through fog or darkness on difficult paths high in the mountains or who in war has had to travel during the night through unfamiliar terrain at the front knows that in doing so all other unpleasant bodily sensations—fatigue, cold, wetness, the heaviness of the rucksack, and the pains proceeding from any injury—can be experienced with an extraordinarily enhanced intensity. The fog need only break up, and all of these troubles disappear almost immediately. ("Whatever was to be done in the autumnal night, the morning has made everything quite better.") No one will want to assume as an explanation for this change of sensations that the optic sensations had inhibited the affective. Fog and darkness are in fact also seen. Thus strictly speaking there is no lack in optic sensations; only the gnostic functions are diminished. In other words, the sensations organize themselves within the total experiencing of an I-environment confrontation. When in the fog orientation ceases, this experience as a whole receives a greater character of suffering and thereby the pathic moment in all individual sensations is strengthened. The same principle is involved in the case cited by Foerster. The laming of cutaneous sensibility does not merely omit individual sensations, and does not merely suspend the physiological inhibition exerted on the affective system by the epicritic system; rather, by the omission of the gnostic factors normally furnished by the cutaneous sensibility, the total experience, even with the same external stimuli, assumes more the character of being-affected and suffering. The same external stimulus is experienced as more penetrating and intrusive. The same stimuli required for the release of affective sensations with an intact epicritic system—stronger pressure and excessive passive stretching—produce not merely sensations of a different quality but experiences with a different sense.

We find an entirely analogous situation in the Freudian example of the experiences mediated by the increasing of sensations. The increasing sensations also mediate the experience of suffering and thus, by virtue of the order in which they appear, have the sense of threat and of danger but only because they are experienced as an increasing of the sensations.

Only when a series of sensations produces the experience of an overwhelming or annihilation do the increasing stimulus magnitudes compel discharge. Our opposition to the genetic theory of anxiety thus waives the empirical objection that the newborn could simply not yet have such temporal orders, directions, and sense-bestowal. We prefer here to emphasize that in every experience of anxiety in the Freudian sense the sensations must be ordered according to the principles indicated here, but that these principles themselves cannot be derived from the individual sensations. The sensations appear as members of a series ordered in a specific manner. Only as members of a series can they be experienced, become intentional objects, and be representative for meaning, as in our example of danger. For the later experiences—the repetitions or reproductions—Freud also assumes in a similar manner that individual sensations and innervations are bound together by a principle of order. He deceives himself, however, when he assumes that in the *first* experience the still unordered sensations could produce the overreaching historical principle out of themselves through the sequence of their occurrence in objective time. We have already shown primarily that the historical formation can unify only because it is itself a uniform multiplicity. And the binding that it brings about must also always be such a unity of multiplicity. This overreaching unity as the ordering principle exists only *by means of*, not *in* the members of the series. Even according to Freud's opinion, the unity into which the sensations and innervations are combined is not made up of a mere objective succession *one after the other*, but rather of a relation *to one another*, in which the members of the unity reciprocally require each other. The sensations and innervations of the anxiety-experience form a unity; this unity is first brought into relief against that which preceded and that which follows by this reciprocal relationship, and thereby creates caesurae in the continuous stream of experiences. *The caesurae arise from the organization, not vice versa.*

One may not object to this polemic that Freud merely wanted to give a rough description of the process and that it is therefore not legitimate to subject the details of his expressions to such sharp and close scrutiny as I have done here. Contrary to this objection, it seems to me that the mode of expression we find fault with here—the concealed sneaking in of concepts strange to the theoretical system—is characteristic at this point not only of Freud, but of genetic theories in general. Coarse descriptions are not enough—certainly not anyway at theoretically decisive points. The indefiniteness of the expression meanwhile is no accidental inadequacy of genetic theory; it is an inadequacy necessary and essential for it, insofar as the goal set for the genetic theories cannot be attained if one proceeds from their genetic presuppositions. Thus the formation one seeks to derive must always be brought covertly into the elements of the derivation.

And yet who would seriously contest that earlier experiences can leave their imprint for later experiences, or that anxiety easily shows up again along with a partial return of the circumstances under which it was once violently aroused—even if the ultimately decisive occasion is lacking. Anxiety-neurotic symptom formations provide numerous illustrations for this. But these phenomena as well as all of the phenomena grouped under the concept of Imago nevertheless have an entirely different meaning than many genetic theories suppose.

Admittedly I have vigorously emphasized that it is the general meanings represented by the individual event that in traumatic and other shocking experiences lift this one concrete process from the series of remaining processes. In contrast to this, however, it cannot escape our attention that memory often preserves events of this kind for us with a livingness and freshness and an abundance of details far surpassing the average. Thus one could infer that it is the particularities of the case that give the total experience its special character. Once we enter into a more exact scrutiny of the facts, however, the experience upon which this objection is based serves only to reconfirm the concept advocated here. We are certain that the individual variations of the sensation themselves play no role in the occurrence of the traumatic experience. Neither the colors nor the lines nor the distribution of light are in themselves frightening or disquieting. The redness, wetness, and viscosity of blood are not capable of having an effect in themself, for example, as this nuance of red or as this gradation of an aggregate condition; rather they call forth an effect in the temperament of the spectator only because they signify blood and because spilled blood itself refers to the metaphysical sense of existence, to life and death. Consider all of the details that could be used to describe the case in a police report—name, age, sex, day, hour, the course of events in the mishap, everything that a photographic plate could retain. These details can certainly not be omitted; yet they can be replaced arbitrarily by other details of the same type without taking away anything of the peculiar nature of the experience. This interchangeability demonstrates that in fact only the general meanings distinguish the individual experience so much above the more unimportant everyday incidents that the experience in its concrete fullness could be even erroneously interpreted as an individual experience, whose obvious *prägnanz*[43] is the decisive factor. The accentuation given to the total experience by the general meanings causes this experience to remain with greater fidelity in memory and to force itself as representative of the general meaning ever again upon the experiencing individual. Genetic theories deal with the origination of the close bond between a specific representative and the act of representation. It is not the origination of experiences that they render intelligible; instead they explain only how that naturally existent interchangeability of all manifest individual factors becomes in many persons so circumscribed that certain experiences can occur

---

43. *prägnanz*: a term in the Gestalt psychology theory of perception. *Prägnanz* is one of the principles of organization in the perceptual field signifying that perceptual configurations that are more clearly articulated, symmetrical, and regular will stand out more vividly for the perceiver. (Translator's note.)

only when bound to specific, prescribed manifest factors. The first experience modifies only the external conditions of its return, whether a few concrete details will later operate as a signal or whether the bond between that which is represented and the representation becomes such a narrow one that the incidental and fortuitous condition of the representing object becomes obligatory for all further experiences of this class. In both groups we are dealing with a specific narrowing-in of the choice among the representing objects and circumstances, that is, we are investigating the harmony between the individual object and experience.

Thus the attempt to reconstruct the situation, in which certain experiences have occurred for the first time, retains a certain justification, namely, to attain information concerning the limits within which such experiences can again occur for the individual. In the remainder of this chapter we will have to verify, using the two most important forms of such processes, signal formation and imago-formation, whether our conception is correct.

Signal-formation is the actual theme of Pavlov's experiments on the formation of conditioned reflexes. Pavlov himself mentions the concept of the signal in the beginning of his comprehensive presentation.[44] Of course he has never attempted an analysis of this phenomenon; rather he has applied it immediately to his experiments in order to derive from them a physiological construction of the mental connection. In this context it has occurred to neither Pavlov nor his critics, that the signal does not merely stand in a relationship with that which is signaled by it (that is, with feeding in *Pavlov*'s experimental procedures); the signal is the middle term of a three-term relation, that is, it signifies the transition from a neutral situation to a nonneutral one. In Pavlov's presentation, reference is made solely to the connection of the signal with the nonneutral situation; he pays absolutely no heed to the other relationship—the dependence of signal formation on the neutral situation. In fact this connection stands out in bold relief at many points in Pavlov's own descriptions of the behavior of his experimental animals, but he gives no attention to this. Had he done so he would have been able to save himself a great expenditure on experiments and would not have found it necessary to fit such numerous auxiliary

---

44. In the following I will cite Pavlov according to the French edition accessible to me: *Lecons sur l'activite du cortex cérébral* (Paris: 1929).

hypotheses into his theory—hypotheses so difficult to reconcile with one another.

Concerning the question, how the signal originates, Pavlov answers, through a physiological, associative connection. As to the further question, what distinguishes the stimulus for the conditioned reflex from the other stimuli of the neutral milieu—one tone from other tones, sounds in general from noises, the entire acoustic sector from the remaining sensory domains, and so forth—Pavlov answers this question too by reference to the associative connection. The repeated presentation of two stimuli, one of which is supposed to release the conditioned and the other the unconditioned reflex, produces through facilitation a binding of the effects of both stimuli. Thus there arises a *new* bond between two physiological elementary-formations of the central nervous system. The facilitation is accompanied in the beginning by the characteristic process of irradiation and later by that of concentration.

What difference would it make if—proceeding from a proper analysis of the signal—we were to consider not only the relationship between signal and what is signaled but also that between signal and neutral situation? If an object is to become the signal, it must fulfill two conditions. It must, even though it is itself neutral (indifferent), nevertheless stand out in relief against the neutral situation. It must be a modification of the preceding condition and at the same time be different in nature from the following specific condition. It is easy to understand that a stimulus applied as a signal must in itself be nonneutral (different). Pop-guns, for example, would not be suitable signals for motorists.[45]

If, however, the signal itself arises from the neutral milieu and is different in nature from the nonneutral condition that follows, then in principal the stimulus applied as a signal must be replaceable by other stimuli. The often heard claim that the conditioned reflexes are simple reactions of the experimental animals to a part of the total situation is a false interpretation of the signal. The signal does not belong to the nonneutral situation as a part but merely points to it. This also follows unequivocally from Pavlov's experiments. The stimuli applied to release the conditioned reflex will have no effect from an arbitrary position, only when they precede the entrance of the nonneutral situa-

45. In idioms such as ''This blow was the signal for a wild brawl,'' the expression ''signal'' is incorrectly used.

tion will they have an effect. If in the training period they always follow the situation, for example, if a certain tone always follows the feeding, then no conditioned reflex will develop to this tone.

Although the signal in itself belongs to the neutral milieu, it must nevertheless stand out in contrast to this milieu in a definite manner. No arbitrary perceptual given and no arbitrary sensory stimulus can be made into a signal by repeated or regular connection with a specific situation. One need only make the effort to think of bad warning signs for autos on the highways. In a certain way broken branches or the arrangement of pavement stones in the pattern of a constellation could take the place of the now familiar signboards: "Dangerous curve," "Unguarded railway-crossing," and so forth. In spite of constant association, however, they would not be suitable as signals because they are not distinct enough from one another and from the preceding surroundings of the driver. They do not modify the situation sufficiently. A conspicuous modification of the preceding situation is therefore one of the conditions for a good signal. Even if according to these conditions no arbitrary stimulus can become a signal, there must nevertheless always be one out of the multitude of equivalent givens that could take the place of the one selected.

How then does this selection corresponding to the nature of the signal, this differentiation of indifferent, neutral material, come about in natural signals that are not created by deliberate arrangement? I would like to illustrate this with an image: There is a meadow that originally can be entered from all sides. One day the owner puts up a fence around his land and leaves only a narrow opening free; now all who wish entrance to the meadow must pass through this opening. At the open place itself absolutely nothing has been altered by the erection of the fence. In addition, no new relation has been established between the fenced-in surface and the surroundings because originally one had been able to get to the meadow by crossing this point, as well as all of the neighboring points. The place left open is the only one in the entire enclosure that has remained unaltered. It only becomes distinct because the borderline to which it belongs has been completely obstructed along the rest of its course. We encounter an entirely analogous situation in the formation of a signal. Originally the transition from a neutral to a nonneutral situation is equally possible at all points. If a stimulus is to become a signal the external circumstances must be ordered so that the transition takes place only at the one point indicated by the stimulus. In order to form a good signal and therefore, in Pavlov's terminology, for a conditioned reflex to be solidly formed, it is necessary that the specific situation enters only when the stimulus selected to be a signal has appeared in the neutral milieu; inversely, as soon as this stimulus shows itself, the nonneutral situation also follows it every time. In Pavlov's experiments if the feeding sometimes follows the tone selected as a stimulus for the conditioned reflex but also follows all other possible events in the neutral milieu, no conditioned reflex will develop. If the feeding frequently follows the conditioned stimulus and only in a small number of cases does not do so, but if the feeding consistently occurs only when the conditioned stimulus has first been given, then in fact a conditioned reflex will be formed, but not as quickly and not as solidly as when the feeding follows the conditioned stimulus every time. Just as the passage to the meadow establishes no new relation, so too the so-called conditioned reflexes are not formed by establishing new connections between the conditioned and the unconditioned stimuli; they are formed only by narrowing-down the possible stimuli of the neutral milieu to one definite stimulus. The Pavlovian dogs must therefore first of all know what's what in a laboratory milieu in order for the process of differentiation to begin. The development of the conditioned reflex is from the beginning nothing more than a process of concentration, that is, of narrowing down and limiting the stimuli.

If we consider Pavlov's experiments from the vantage point supplied by the analysis of the signal, then we may succeed—easily and in a simple fashion—in clarifying phenomena that *forced* Pavlov to assume and assert a series of complicated hypotheses.

Let us begin with the initial situation. Pavlov himself gives a number of examples showing that it must be a neutral situation. Among others he mentions a change in the experimenter or in the laboratory space, an unforeseen invading tone, a sudden change in illumination such as the disappearance of the sun behind the clouds, and a draught of air invading from the door and laden with smells of some kind; any one of these will suffice to hinder the function of the conditioned reflex.[46]

Pavlov interprets this disturbance as the effect of inhibitions proceeding from "trace-reflexes" released by the new stimuli. The one-after-the-other succession of initial situation and transition situation

46. Pavlov, pp. 45–46.

becomes for Pavlov a side-by-side juxtaposition of the conditioned reflexes and the superimposed inhibiting reflexes. He fails to see that a new and nonneutral initial situation has been created. The signals for the conditioned reflex, however, can work only when there is no gradient[47] from the initial situation to the signal. The initial situation and signal must lie on the same level and thus be neutral to one another. Pavlov's book also contains examples showing that the signal must be neutral; an electrical current applied as the stimulus for the conditioned reflex no longer functioned as soon as a certain intensity was surpassed. In this case the stimulus establishes a self-contained experience, which cannot refer to another. In the reaction occasioned by an intensive pain stimulus—that is, the reaction of getting away from the source of the stimulation—the experiencing is backwards, turned toward the past. When the pain is reduced a caesura occurs, closing off the experience that began when the current was switched on. The caesura hinders the stimulus function of pointing into the future, that is, its function as a signal, as a stimulus for the conditioned reflex. To explain this experience Pavlov was forced once again to refer back to the assumption of the physiological processes of inhibition.

We do not aim here at an exhaustive confrontation with Pavlovian theory. Our concern at the moment is only with presenting the process of *circumscription*, which serves as the basis for the development of conditioned reflexes. I do not wish to go into all of the difficulties that Pavlov encountered from the beginning—difficulties resulting simply because he overlooked the phenomenal structure of the signal. It is enough for us to show that in spite of all opposing theoretical interpretations the three-term organization characteristic for a signal can be clearly inferred from Pavlov's own descriptions of the experimental conditions.

On the other hand, we can most easily uncover the process of circumscription in Pavlov's portrayal of the generalization of a conditioned reflex and in his theoretical construction of its irradiation and concentration.

During the development of a conditioned reflex, in a certain stage of the training, the experimental animals will react to skin stimulation no matter where on the body surface the stimulus is applied or will react to tones regardless of tone-pitch, tone strength, timbre, or the nature of the instrument forming the sound, and so forth. The reactions will still

47. This expression should not be taken in the physiological sense.

occur, even if the *one* single tone, such as D^2, or the stimulation of a single particular place on the skin has been used to generate the conditioned reflex.[48] If the training is carried far enough a further differentiation of the tone-region is gradually achieved: the experimental dog receives food only to a certain tone, while after other tones newly introduced into the experiment it obtains no nourishment. After sufficiently frequent repetition the animal will react with salivary secretion only to the one desired tone, while with the others no more secretion will occur. Pavlov explains the initial general reaction to stimuli of a certain class and the subsequent narrowing to a single variety; he attributes this to an initial irradiation of the stimulus effect to neighboring parts of the cortex, followed by a decline in the irradiation, and later by a concentration.

For Pavlov it is self-evident that the one tone used in the experiment can have an effect only as this one, specific tone. The idea that a tone comes to be experienced first as a tone in general—as an alteration of the environment through sound—did not enter his mental horizons. In this respect his *mosaic-theory* and the *constancy hypothesis* hindered him.[49] If, however, the tone nevertheless operates unspecifically in the beginning of the training, Pavlov can only explain this as a co-excitation of other specific analyzers. The wave of excitation must reach beyond the initially excited analyzer, which is specifically associated with the chosen tonal stimulation, to the related neighboring brain elements, and thus dispose them for the conditioned reflex.

Nevertheless, if any kind of sensory stimulus were to become the signal, both through the modification and narrowing-in of the preceding situation and through the constant connection with the following, a special set of circumstances would be required in order to bring the experimental animal to further differentiation. Even in the monotonous environment of a laboratory the offering of food can still be

48. Pavlov, p. 105.
49. Pavlov assumed that there is a specific apparatus in the nervous system corresponding to a specific incoming stimulation. Such an apparatus, called an analyzer, serves as the basis for stimulus selectivity, according to Pavlov. Thus an indiscriminate reaction by the organism to several different stimuli in the same class, such as several tones of different frequency, was explained by the generalization to, and involvement of, neighboring analyzers, or irradiation. These hypothetical analyzers were thought to consist of peripheral endings of sense organs, the afferent nerves pertaining to them, and the endings of these nerves in the brain. Thus Pavlov's model of the nervous system and its functioning is a *mosaic* of discrete individual apparatuses and functions (*mosaic theory*). He also assumes a *constant* relation between one specific incoming stimulus and a specific apparatus in the nervous system with its specific sensation (*constancy hypothesis*). (Translator's note.)

preceded by such a multiplicity of sensory stimuli—such as smells, noises, change of illumination, coming and going, and the talk of humans—that the regular sounding of a whistle or small bell presents an adequate modification for signal formation. The filling of the environment with sound in general is the signal during the first segment of the action. If the dog is now to be further trained to a specific tone, the feeding must be omitted after all tones other than the chosen one. This alters absolutely nothing in the constancy of the coordination with the tone used from the beginning. This tone and the feeding have from the beginning consistently taken place together.

There is thus no need to assume that the process of irradiation changes suddenly into the process of concentration. Training to a specific tone is the direct continuation of the preceding training. Just as now the feeding follows only after a single tone, so indeed from the beginning the feeding has folowed only after tone in general but not after smells, strokes, or talking. The signal is thus from the outset a modification of the environment and narrowing-in of the possible stimuli, first to a certain sensory sphere and last to a single, specific stimulus of this sphere. The reaction of the experimental animal to the sphere of sound, that is, that stage that Pavlov characterizes as generalization, is an unavoidable stage-in-transit in the training. As exactly as possible Pavlov observes what objectively happens. In doing so, however, he does not notice everything that he omits doing in that he does just this one thing.

An animal that is kept in persistent noise can be trained equally well to a suddenly commencing stillness as a signal for the coming specific situation. The Pavlovian school can explain a conditioned reflex of this kind—released by the suddenly commencing stillness and thus characterized by the lack of physical stimuli—only by a series of auxiliary hypotheses. If we simply proceed from the nature of the signal, however, then we will have no difficulty in understanding the stillness as a distinct modification of the preceding noise-filled environment. The stillness can serve just as well as the sound, the darkness just as well as the light, to fulfill the one requirement essential for the signal, that is, the modification of the environment.

Pavlov reports that dogs brought into the investigation-room while the rhythmic noise of a metronome was audible could be trained so that the salivary secretion of the conditioned reflex set in as soon as the noise of the metronome was interrupted. Even the sudden diminution of the tone intensity sufficed for the same effect. But if the decrease was undertaken very gradually it had no effect. The suddenly discontinued trumpet tone D^2 (high D) produced a conditioned reflex to the measure of 32 drops of saliva per minute. The same tone, gradually decreased in the course of 12 minutes until it disappeared entirely, was without effect.[50] This observation placed Pavlov into a singular embarrassment, and he knew only one way to extricate himself: he introduced the concept of trace reflexes. The stimulus for the conditioned reflex is supposedly not the tone used, but rather its trace in the cortical level of the central nervous system. This contrived explanation becomes superfluous as soon as one properly considers the nature of the signal and the process of circumscription, which is so decisive for signal formation, and as soon as one replaces the constructive, physiological theory with a psychological theory based upon a phenomenological analysis of the signal.

When we have once done so, then the experimental conditions just mentioned will clearly tell us that it is not the material substratum of the stimulus that has been operative—because the stillness in fact has nothing of that nature—but rather the phenomenon of *transition*. But for Pavlov temporal becoming is not comprehensible; his theory is based upon the dissolution of the nervous system into a group of spatially separated analyzers, and this dissolution forces him also to assume a point-by-point temporal succession.

First one analyzer commences activity, then another, and finally a third, in an order determined by the external stimuli. No inner relationship of any kind can exist within this succession of events one after the other.

When we survey the work of Pavlov, when we call to mind the monumental effort applied in order to construct an explanation of mental processes according to a mechanical schema, and call to mind as well all of the experimental designs modified again and again and the proud protocols of objective observation, then a dreadful thought forces itself upon us. We could well imagine that someone—lacking any kind of knowledge of electricity and its manifestations—had once observed that an electrical car goes into motion following certain manipulations of its driver, that other acts of the driver accelerate the car or brake it and bring it to a standstill, and that this person had

50. Pavlov, p. 39 ff. (French ed.).

attempted to penetrate the mystery of electrical power by an exact plotting of these movements. With an enormous amount of precision equipment he seats himself next to the driver and records with the most painstaking exactitude his individual manipulations and the positive and negative acceleration that they impart to the car. He thinks out a theoretical construction in which the movements of the driver are augmented as operations of leverage on a gigantic scale and now believes himself to have recognized how the car is propelled. On the other hand, he indignantly refuses to take a look into the interior of the car or to accept instructions concerning the nature of electricity, the construction of dynamos, or the conduction of electrical current.

Pavlov's entire attitude obstructs him from comprehending the temporal organization of mental experiencing, from becoming aware of the phenomena of retention and anticipation, or from ascribing any kind of meanings to them. His theory distorts the temporally ordered, three-term relation in which the signal stands as middle term. First of all the initial situation is merged with the transition situation into a conglomeration of different kinds of reflexes. Once this has happened, the signal can also no longer be a reference, that is, an anticipation of the coming different situation. In Pavlov's system there is only the immediate material contact of cause and effect, touching in a point in time. The animated organism's anticipation of what is coming and its reaction to it has no place in his theory. There the conditioned reflex with its stimulus effect takes the place, even the temporal place, of the absolute reflex. According to Pavlov's basic concept there can be only reactions of the organism to what is objectively present. He therefore fully overlooks that the conditioned reflexes involve not the representation of the absolute stimuli by the conditioned stimuli but reactions to the approach of a different situation, that is, reactions to what is in the future. In the last analysis the impossibility of incorporating the temporal organization into his system hinders Pavlov from recognizing the ontological structure of the signal and from noting the process of circumscription by which the signal is formed.

We need not go any further with the investigation of the problems touched on by the theory of the conditioned reflex. It has now become comprehensible how a particular given, which is not an essential component of an experience, can nevertheless—through the process of circumscription going on from the earlier experiences to the present— become an important and under certain circumstances even practically indispensible condition for the occurrence of certain kinds of other experiences. Thus we have answered the question raised above concerning the connection of the earlier with the later experiences, though only in part; that is, we have answered only the question regarding those instances in which one given points to another given that is strange to it but is from the same sphere of reality; we have not answered for the instances in which a manifest given represents a general essence.

And yet this group—in which the accidental, particular qualities of the representing object have become an indispensible condition for its representative functioning—is for our purposes the more important one. Perhaps we can make use of the results of the preceding investigation here too by allowing ourselves the assumption that the process of narrowing-in could be decisively important for this group as well.

For instance, consider the case in which the erotic choice of the young man is secretly determined by the image of the mother. It would follow from our assumption that in the beloved as well as in the wife the young man does not once again seek the mother as this certain individual personality; rather the beloved becomes a likeness of the mother, since ultimately both are likenesses, the mother no less than the beloved—likenesses and embodiments of the essence of "the Motherly" and of "the Feminine" in general. The first impressions that children receive of the mother fill up and dominate them so that the peculiar individuality of the mother, her voice, her smile, her gait—all of the thousand individual features that, changing from woman to woman, give to each her special imprint—will be experienced in the future as the only possible expression of the feminine. This prototype of the woman was drawn after the individual appearance of the mother. It is to this prototype, as its most perfect actualization, that the choice now turns.[51]

Wherever the human being has to make a decision important for him, wherever he can choose and does choose, the object in its

51. We cannot here pass over C. G. Jung's teaching of the collective unconscious or of the archetypal images because we meet with the same misunderstandings there as in other genetic theories. Characteristic for this are Jung's comprehensive statements in his *Psychological Types*. There it states:

> From the scientific, causal standpoint, the primordial image can be conceived as a mnemic deposit, an imprint or *engramm* (Semon), which has arisen through the condensation of countless processes of a similar kind. In this respect it is a precipitate and, therefore, a typical basic form, of certain ever-recurring psychic experiences. . . . From this standpoint it is a psychic expression of the physiological and anatomical disposition. If one holds the view that a particular anatomical structure is a product of environmental conditions working on living matter, then the primordial image, in its constant and universal distribution, would be the product of

individual particularity becomes the matter of the choice. When the choice narrows down to the final individual nuances, does the choice then also, as appearances could lead us to presume, concern these last particulars themselves? When, as usually happens, a man becomes "picky" with age, does that mean somehow that in all things, even the most decisive, he has oriented himself entirely toward the particular and the peripheral? Or has he merely learned better to distinguish and to order the individual objects according to their particular quality in their representative function? Whether someone prefers blondes, brunettes, or black-haired women, nevertheless the color as color never determines his preference. The black, which attracts him as the color of hair, might leave him entirely cold as the color of a cloth, of a piece of furniture, or of a roof. On the other hand, if it should be objected that naturally it is not the color that is meant—not the blackness of the hair but rather the black hair—then we need only place the counter question, What then becomes of the magical power of black hair, as soon it lies in a carton processed somehow into a wig? Or, to take another example, the arch of the eye brows, the rounding of a shoulder, what would the same lines have to signify as ornament, as geometric curve, or as contour of a machine part?[52] One need only

---

equally constant and universal influences from without, which must, therefore, act like a natural law. (C. G. Jung, *Psychological Types*, trans. H. G. Baynes; rev. R. F. C. Hull, Bollingen Series XX [Princeton: Princeton University Press, 1971], pp. 443–444.)

Jung obviously qualifies this statement further:

Accordingly, the primordial image is related just as much to certain palpable, self-perpetuating, and continually operative natural processes as it is to certain inner determinants of psychic life and life in general. The organism confronts light with a new structure, the eye, and the psyche confronts the natural process with a symbolic image, which apprehends it in the same way as the eye catches the light. And just as the eye bears witness to the peculiar and spontaneous creative activity of living matter, the primordial image expresses the unique and unconditional creative power of the psyche (p. 444).

By this interpretatoin Jung throws aside his own principle, which he had just previously declared; he doesn't see that the "natural process," as we have presented it above for birth, is already the product of an ordering mind. Unfortunately Jung does not disclose how something can be simultaneously conditioned and unconditioned. In the context of the history of ideas, Jung's teaching of the archetypal images is merely a belated effort to resolve the problem of universals psychologistically.

52. I am obliged here to fulfill a literary responsibility. After this section was completely written, I was surprised by a concurrence at this point with Cassirer's *Philosophy of Symbolic Forms III*, and at a later point with Heidegger's brief text "What is Metaphysics?" I have been forced to postpone the study of Heidegger's large work *Being and Time*. Although the delight in being able to refer to significant authors suffers when one is dealing with works that have just now appeared, nevertheless a reference to the texts mentioned should not be omitted.

raise these questions to see at once that that which is sensually evident does not have an effect as sensation but as expression. But as expression of what? From the blackness of the hair we have come to the black hair, and if we now question further in a like manner we will always be referred from one part of the bodily appearance to another and will nevertheless in the end find in none of them our true objective, which expresses itself in the color of the hair and in the rounding of the shoulder. The essence is actualized in the individual person and is embodied in his or her external appearance, but this essence cannot be grasped entirely in any single part. The whole shines through that which a moment in time is capable of presenting; but no matter how comprehensive this may be, it only shines through and is never spent fully in it. Each instance is only one moment from the life of a man. In none is the man fully contained; rather every moment gives only a more or less distinct or powerful expression of this totality without fully exhuasting it. The individual is never entirely given in the moment elapsing in time, even to himself. Since everything, which the moment shows, is only an expressive part of a whole, since, for example, even the totality of the bodily appearance is only a part, any other more narrowly delimited part whatsoever such as a lock of hair or an image can also receive the value of the relic. It is not as remains of their living organisms that the bones of the saints are the support for remembrance and veneration; both—the bones and the body to which they once belonged—are manifestations of the person of the saint.

No psychological explanation is adequate or necessary for this relationship of the essence to its manifestation. Psychology encounters it as a fact. Psychologically we can explain only why, in the individual case, this or that detail of the phenomenon as experienced is to a greater extent representative. In distinguishing one detail before many others of the same kind and worth, which likewise could be possible carriers of the expression, we recognize, along with other factors, the searched for aftereffects of earlier experiences. That somehow the exact appearance of the mother has become the model, and that her characteristic individuality determines the prototype of the woman—all of this may be derived from the constellation of circumstances. The individual who has become selective through his own experience has learned with increasing certainty to distinguish expressive value. Thus with the full development of his capacity for experience, the perceptual givens are ordered for him into groups according to their expressive value, and his choice is directed toward the most highly valued among them. However, for the anaclitic type the appearance and essence have from childhood so firmly coalesced with one another, that the capacity for experience has thereby been diminished. He perceives expressive value generally only in one or a few groups among the many possible; these then become for him the solitary representatives of the general essence. But even in this limitation the choice is still directed, through the particular detail of the appearance, toward the essence manifested in it.

To choose the particular detail for that particular detail's own sake—this is the affair of the refined. In refinement, therefore, the choice remains one—one of many. Inconstancy of relations and the partial character of the being-affected and surrendering are therefore necessarily and from the outset properties of refinement. It would be remarkable if the individual distinctions on which their choices are ultimately based had the same meaning for the lover and the refined individual. While for the refined individual the one chosen is simply one out of many that remain enticing and desirable alongside this one, in the love-filled bond the attraction of all others disappears beside the image of the beloved. To the lover the beloved appears as the unique, unforgettable, and irreplaceable. Thus the erotic relation can be distinguished from the refined fascination with the particular, not only by the intensity but also by the extent and the depth of the being-affected. The refined individual enters into a relation only playfully, ''for the time being'' as it were, and only with a part of his person; the lover on the other hand is entirely affected.

The erotic decision and the refined choice are thus both disposed toward the particular detail. But the details in the two cases belong to entirely different orders of being even in the event that the two different kinds of choosing—the refined and the deeply-affected—aim at the same manifest quality of one and the same person.

4 / *The Thematic Content*

Childhood impressions and shocking experiences open one's vision for certain general givens and close it for others; they thereby create a principle of selection, and consequently some situations and objects acquire a preeminence through their representative function, while others undergo rejection. The aftereffect of childhood impressions is detectable in every individual without exception, and without prejudice to his constitutional individuality. This generality corresponds to the historical modality of the childhood impressions, their position in the beginning of individual development. If, on the other hand, we examine a group of youths or adults for their receptiveness to shocking experiences, then already from the early years on, soon after the close of infancy, we find remarkable differences. In fact our general assumption is that the same situation that one individual endures calmly, will move another in the most profound way.

We may conclude from the first-time-ness—which we have reckoned among the structural moments of the shocking experience—that the objective lack of experience, that is, the untouchedness, is one of the presuppositions for the occurrence of such experiences. Nevertheless this is only an exclusive and negative presupposition not a sufficient and positive one. Two men may share the same objective lack of experiences, the same untouchedness, yet the same situations need by no means call forth the same shock in both. Two men of the same age, who have grown up in the same narrow rural environment, can nevertheless be absolutely different in susceptibility for shocks, even when the same environment has presented them both with entirely similar impressions. This susceptibility has nothing to do with the vulnerability of the sensitive. In the following discussion, for the sake of clear terminological distinction, we will call this susceptibility the *subjective readiness*, and by this reference simultaneously point to the fact that it contains a researchable, intentional content, that it is readiness *for something*. We can thus discuss the subjective readiness only by simultaneously going into the *thematic content* of the shocking experience.

The comparison of the deeply-affected with the refined mode of choosing at the close of the preceding chapter has cleared the way for this task. In the "being-entirely-affected" we confront a phenomenon in which we can comprehend how the subject experiences himself in his experiences. We may conclude from the contrast between the "being-entirely-affected" by the shocking experience and the superficial being-touched that not only has the object a representative function but also the single moment of the stream of experience is experienced as representative, that is, representative for the whole of the individual person. The experiencing individual *has* himself in single successive moments, as it were, in adumbrations (*Abschattungen*) to use Husserl's term. Each of these views presents the individual whole of the person in different clarity and fullness.[53]

As is so common in psychology, we must have recourse here to spatial comparisons and expressions. But even if, in lieu of clarity, we had used other expressions not borrowed from the sphere of the visible, all such words remain equally misleading and ambiguous for anyone who does not want to, or cannot, acknowledge the moment of representation. We also encounter a special difficulty that directly hinders the advancement of the psychology of feelings, that is, in this area the spatial schema breaks down so readily. For the object-world, space easily becomes the representative of duration, of the thing, of being, of the whole. How then are we to conceive the whole of the individual person, which presents or actualizes itself in the single moments of experience? We must forgo an exhaustive answer to this question and must suffice ourselves at this time with pointing to the phenomenon of "being-entirely-affected." We may clearly conclude from this

53. The comparison of the refined and deeply affected modes of choosing thus discloses an important perspective for us: it makes possible a deeper insight into the structure of feelings. The relationship of the moment to the whole of the person appears to me to be the essential intentional content of those kinds of experiences that we characterize as feelings. It should not be difficult to recognize that this concept differs in principle from that of Krueger, despite the terminological similarities. (F. Krueger was a Gestalt psychologist and author of several works including *Qüalitaten, Gestalten, und Gefühle* [*Qualities, Gestalts and Feelings*], 1926). (Translator's note.)

phenomenon that the subject not only experiences *something* in the experience but also at the same time *himself*. The symbolic relationship of the moment to the whole of the person, the subjective representation, ranks along with or takes precedence before the objective representation. This whole is concretized and actualized in the historical figure of a human being. In every single moment of his life the human being works at the actualization of this figure; every single moment explicates the individual whole, and guides it more and more out of a potential and into an actual being. Every single moment is subordinated to the whole; the whole exactingly governs the formation of the single moment. In feeling, this relation of the single moment to the whole of the personal existence is directly, immediately comprehended originally in the form of the object, upon which the transition from potential to actual being—that is, the process of Gestalt-creation—is carried out. As soon as the analysis of consciousness is no longer—as has most frequently been the case up to now—subordinated to epistemological aims, and as soon as "historical reason"[54] takes the place of pure reason, it will also be possible to rigorously found these theses.

In the same measure and with the same step in which he gets "to the world," the experiencing individual also comes for the first time to himself. It has often been observed and often said that the whole of the world can be intended in an action, even if the world were actually conceived or taken hold of only in a part. Just as the part taken hold of is representative for the "whole" of the world, so too the experience concluded in one transient moment is representative for the whole of the person. The ready subject is directed toward this whole of subjectivity, which is concretized and acquires its content in the instant of the shocking experience. Thus the depth to which a real object will become transparent for the view of the experiencing individual depends on subjective readiness.

Subjective readiness is conditioned by hereditary predispositions (*Anlagen*) as well as by the inner and outer life history. Even in cases in which, because of hereditary predisposition, the capacity for concentrated experiencing is strong, the subjective readiness varies, following

54. This expression was first used by Wilhelm Dilthey in his *Einleitung in die Geisteswissenschaften* (*Introduction to the Human Sciences*) (Leipzig: 1883). W. Strich has attempted to present the sense of this expression concretely in his valuable *Prinzipien der psychologischen Erkenntnis* (*Principles of Psychological Knowledge*).

the ups and downs of biological events and the course and the sense of life history. The curve of this longing and this satiation is entirely different from that of instinctual desiring and instinctual satisfaction. Even midst the full satisfaction of instinctual drives, the subject who is ready for fulfillment can remain unfulfilled.

Fulfillment is bound to definite objective situations and objects. If the suitable objective situation were lacking, then the subjective readiness would make itself known in yearning. Yearning in this sense is not longing for the renewed presence of an already familiar man, thing, condition, or landscape; rather it turns out to be a yearning after something unknown, a longing to come entirely to oneself. This distinguishes yearning from carnal desire, which according to its meaning already has its object. Just as the yearning of the subjectively ready can find its fulfillment only in a suitable object, so too when the subjective readiness is lacking, no matter what object may appear in the visual field, no shocking experience will occur.

Of course we will have to further limit the general validity of the last thesis as we inspect more closely how the subjective representation, that is, the same thematic content, conditions different modes of behavior according to the degree of subjective readiness. It attracts one individual toward certain matters and configurations yet repels another from them and leads to shock or to psychic trauma.

If we are seeking phenomena in which the subjective representation, the relation of the single moment to the whole, attains the most comprehensible expression possible, then we must consider the ''limit'' situations (Jaspers)—experiences such as the nearness of death and the readiness for death—as preeminent. In the readiness for death or even the blissfulness at death that is shown by lovers, death is not the nullification but rather the consummation of life. In many cases of lovers sharing death, the rational motives will often not appear weighty enough to an outside observer to account for their deed; however, these reasons win power over the lovers as easily as they do because they make it possible for them to actualize the immanent meaning of their experiencing. Life is not surrendered because it no longer carries meaning or because it has become unbearable, but rather because in this one moment the whole seems to be completed. The *one* moment, lifted out of a succession of moments perishing in the stream of time acquires an absolute importance and meaning. The deed whose motives are opaque in the sphere of reality becomes comprehensible as the expression and objective-formation of an experience.

Nearness to death also gives its special imprint to bold undertakings, to danger, and to adventure. The enticing quality of adventure, however, cannot be explained simply by the tension that arises from the movement one step from the threshold of death. Anyone who was forced in the war to undergo a long bombardment noticed nothing during these hours of the magic of the bold venture—in spite of the nearness of death. It was much more a dull enduring, and a torturing experience of powerlessness at being handed over to a senseless, blind accident. The mountain climber, on the other hand, who is perhaps in no less danger for hours on a difficult path, experiences every hand-hold and step, every ascent and descent as a wonderful expansion of the limits of his existence. The nearness of death is a condition for such an experience. Only at the threshold of death can the limits of life be carried forward. The forbidden thus entices just like the dangerous. It is not only defiant rebellion against the proclaimer of the prohibition that drives one to transgress against it. There are in fact sufficient cases in which prohibitions have the character of a warning. It is precisely in these cases that it becomes clear that in the transgression the expansion of personal limits is experienced and one's own powers are tested. In adventure, death is put to the test. Death is not sought; it can occur, but it is not supposed to. It does not belong inherently to the experience as in the case of romantic love, for which death is the presentation or the actualization of the sense of the experience.[55]

That dying should be an actualization of experiencing sounds strange, perhaps even contradictory. Such paradoxes appear regularly wherever the attempt is made to express the absolute by means belonging absolutely to the sphere of what occurs, elapses, and passes. Human life is full of phenomena of this kind; they remain incomprehensible as long as one does not perceive the sense that encroaches on the place and the time boundedness of their natural existence.

Even the artist or the researcher must serve a natural medium—of

---

55. These references to love-deaths as the actualization of romantic love may appear peculiar to the reader. The kind of romantic love epitomized in German romantic literature and operas, however, receives its most perfect fulfillment in death, as is evidenced by the death of one or both lovers in the final scenes of *The Flying Dutchman, Tristan and Isolde, Tannhauser,* and many others. We need only think of *Romeo and Juliet* and *Othello* to realize that love-death is not a uniquely German phenomenon. (Translator's note.)

tone, of color, or of word. But his work is identical neither to the material from which it is composed, nor to the formed material of which it consists. By the Ninth Symphony we do not mean the manuscript preserved in the Berlin library, but rather the creation of Beethoven, which is fixed only in the written copy and is reproducible in any number of performances. The manuscript can burn up, the notes of the concert can fade away, but the work remains in existence. Even in the plastic arts the work is not the object that the art dealer buys or sells. The connection to the material is certainly a more intimate one there than in music, poetry, or the sciences. Nevertheless there is no identity between work and material, not even in architecture where the connection is the most narrow, where in practice a reproduction is usually not possible. In the same way the sense of the romantic love-death is the consummation and not the termination of life. While here death itself becomes the real representative of the whole and belongs to the already completed experience of subjective representation, in the bold undertaking only the *nearness* of death is the condition for the possibility of an intensified experience. While every bold venture, of whatever kind it may be, plays itself out in the nearness of death, of solitude, and of immediateness, a shocking experience cannot be sought out without a subjective readiness to tarry in such situations. Even instinctually driven action can come into the nearness of death; but it is not sought as it is in romantic love; it is also not withstood or overcome as it is in the heroic experience. The driven individual falls into the nearness of death as the objective consequence of his actions, and in spite of his tendency directed toward another goal. Instinctually driven action is present, repeatable—a mere event; it is the authentic antithesis to a nonrepeatable, representative-historical experience.

In itself each actualization of the whole remains merely a representative, that is, the whole itself transcends the experience. Experiencing pushes out of transitoriness toward self-actualization, toward the creation of its own form and of the work. The realm of timeless being, which the individual touches in the moment of the deed, disappears from him as soon as the work has come into being and the deed has been accomplished. He remains imprisoned by life and by time. Just as the form is exposed to perishing and the work to destruction, so too the creative individual—already in the moment of consummation—sinks back again into the flow of everyday events. His work leaves him behind; he cannot hold it. We are concerned here not with the fact that the work, as soon as it is created, acquires an independent existence, has its own fate, and outlasts the creator. This depends on many circumstances, not least of all on the maturity and perfection of the work. It is inevitable that the creator outlives his work, that is, in the moment in which he completes the work he has already lost it. The good fortune of accomplishment and concentrated experiencing are followed by collapse. The whole of his own being, which he was close to in the one moment, vanishes again and leaves him in a state of doubt, of emptiness, and of disappointment. For this reason many drag out and delay the conclusion. For this reason each completion places the individual before the necessity of a new beginning. The hubris that drove the heroes of the tragedies of antiquity to their downfall is inner destiny. Because it unveils the true nature of the man, the tragedy shocks, and fear and sympathy are its effect on the spectator. The catharsis effected by it is not purification from the great passions, as Aristotle taught, but rather from the small scruples and doubts of everyday life.

In modern history this compulsion requiring one always to begin anew and to reach ever further to the point of one's own destruction, has found an imperishable expression in the phenomenon of Napoleon. No goal satisfies him any longer, once he has arrived at it. In forging ahead he arrives at one moment at the outer limit of what is possible, he arrives at the point in which the whole seems to be represented. But immediately, at the outer limit of what has been attained, the new realm of the possible presents itself and the whole recedes before him, just as the wanderer moves the horizon forward with every step he takes. The period of the Consulate, the brief period of satisfaction and order in France, is followed by the expeditions of conquest, which are supposed to establish the world-empire.[56] The world, the whole, which at every border appears incomplete in new distances, summons him forth. Others, individuals and whole peoples, have gone the same way in the century-long duration of their existence. And all have made the same discovery, that beyond every goal a new one comes forth, that the limits of that which is possible wander with them. The passionate striving for the New, not for the sensation but rather for the ultimately

56. Napoleon overthrew the Republican Directory in November 1799 (18 and 19 Brumaire in the year VIII by the new calendar of the French Revolution), and established himself as First Consul in a consulate of three. (Translator's note.)

fulfilling New, is characteristic of the inventor, the researcher, and, with a certain difference, the sportsman, the political man, and the conquerer.[57]

Someone living today necessarily takes different tasks upon himself than had he undertaken them one hundred or two hundred years ago. The vehemence, the perseverance, and the fearlessness with which he holds fast to his problem should not blind us to the fact that these physical or medical questions, this uninvestigated region, or this state resisting subjugation, are nevertheless not ultimately intended in themselves. They are, in any case, intended only insofar as they signify something *beyond* the limit and at the same time also represent, in whatever gradations, the *whole* of the world. "Beyondness" and wholeness (totality) define the tasks an individual chooses for himself and determine the direction of his experiences from the side of the object (task-object). As concerns the subject, what is decisive is not only talent and giftedness but also the courage and strength to sustain himself in limit situations and to move through them. This formulation of the limit situation completes our effort to sketch out the mode of the concentrated experience for the first time in its rough outlines. It nevertheless continues to make a decisive difference, what kind of limits a man approaches and what kind of forces he risks provoking.

The expeditions of Alexander, Barbarossa, and Napolean are world-yielding, as are the first ascent of the Matterhorn and Lindbergh's flight over the ocean. In the latter cases, the mountain and ocean, and in the earlier countries and peoples, serve as the material forming the experience. The greatness of the material on which someone tests himself and in which he seeks himself are determined just as much by the greatness and breadth of his own nature as by the historical place at which he finds himself. We mention the great names because the phenomenon of subjective readiness, the determination of the course of life by the *existentiell* content of experience, come most distinctly into view in those bearing these names. In others this deter-

---

57. The unstable individual falls prey to the charm of the New insofar as it is something unusual, unaccustomed, and different from anything present earlier. His experiencing is almost entirely lacking in the subjective representation of the moment and the objective representation of the object. His life disintegrates for him into a succession of detached moments, just as the world disintegrates into a conglomeration of singular objects. The weakness of the representative function has as its consequence the directionlessness of the unstable individual, the continuous fluctuation from one goal to another. On the other hand, in the striving after the fulfilling New, the one direction is maintained throughout all of the changes in the objects.

mination is definitely more difficult to demonstrate, but it is nevertheless by no means lacking. On the contrary, every experience and every
structuring of life viewed as a whole takes up a position toward and
answers the *existentiell* question that continuously frames every experiencing. This is not a theoretical answering; it is an answer through
the deed. Even everyday life still offers material in which the ready
subject can actualize himself. Yet the material becomes ever more
resistant and difficult so that not many succeed any longer in objectifying themselves in it. They search for and reach after a substitute such as
that offered by adventure, gambling, and daring sporting achievements. Even the enthusiasm in the year 1914 probably had its ultimate
source in the newly awakened hope for the actualization of *existentiell*
meaning.

In the gambler we already meet up with a distorting feature of
passivity; to be sure he still risks provoking fate but in doing so he no
longer confronts it as an actor forming his own fate; rather he simply
accepts its decision, whether it be bounty or annihilation. A few steps
lower we meet up with those who dare only to encounter *existentiell*
questions as spectators of others' fates, who suffice themselves with the
horror available to a spectator. The powerful attraction of the variety
theater and the charm of dangerous acrobatic arts prove how widely
diffused this inclination is. From here we have only one step down to
the life-formation of the man whose basic principle is holding at a
distance *existentiell* questions—the bourgeois as philistine (*Spiess-
bürger*). We may infer, from the quantity of preparations he makes in
order to hold the invasion of the *existentiell* question at a distance and
from his industrious activity how difficult it is to screen himself against
it and to defend himself successfully before it. At the end of this series
stands the phobic who breaks down in anxiety if he merely catches
sight of the *existentiell* in the spatial qualities[58] of width, depth, and
narrowness.

Thus while at the beginning of the series developed here we find the
heroic man with a marked subjective readiness for a decisive, formative response to the *existentiell* meaning of experience, at the end of the
series we meet the phobic who flees the absolute and the limit situation.[59]

58. We must reserve an analysis of this concept of spatial qualities and of its significance for
the understanding of the phobias, for a separate publication, although you may also wish to
consult the relevant comments on p. 80 ff.

59. I am quite conscious of the gross schematism of this series. If one lets the "gambler" and

The phobias remain incomprehensible as long as we search for a naturalistic explanation. They become comprehensible as soon as we cease our resistance against becoming aware of their *existentiell* meaning. But it is not only the phobias themselves that become comprehensible. Their analysis, which we cannot undertake here, would simultaneously perform the practical service of definitively demonstrating the relatedness of all experiencing to *existentiell* meaning.

In determining the *existentiell* content of experiences we nevertheless still draw absolutely no conclusions regarding their knowledge-value. We leave entirely undecided whether or not a metaphysical truth is adequately grasped in the *existentiell* content of the experiences. After all we are pursuing psychology and not epistemology. Therefore critical reflections concerning the possibility of metaphysics in general cannot validly be used against our position. Such reflections address themselves only to the question of whether metaphysics may raise the claim to be scientific knowledge. Like all scientific theses, the teachings of psychology make claims to truth, but *erring*, being in error, belongs to psychology's subject matter as a basic feature of human nature. Formal logic investigates only the structures and the relationships of (correct) thoughts, the error for it only settles the boundaries of logic; the error is not a problem for formal logic. In contrast, the process of thinking, of erring as well as of correct thinking, is an object for psychological study, and the question concerning the possibility of error is a basic problem of theoretical psychology. This signifies a considerable expansion of the sphere of problems in comparison to formal logic. The possibility of error is not eliminated by the most precise knowledge of logical theses, just as in general the phenomenal conduct of the human being in its basic features cannot be altered by scientific findings or knowledge. The physicist does not move about in a world of waves, electrons, and electrical fields; he lives, as do we all, in a world of tones, colors, and things. No more so do any kind of

---

the "phobic" appear as types, then all of the general objections raised against typology are valid in their entirety. The great inadequacy of typology is that the single modes of behavior are so distributed among definite figures, as though only these modes of behavior were encountered in them and these modes continuously and permanently. The gambler, however, is no more *only* a gambler and the phobic is no more only a phobic than the syntonic is only syntonic and the schizoid only schizoid. In spite of the legitimacy of such objections the typology remains a legitimate device for presentation and an unobjectionable one so long as one takes it for nothing more than a schema for guiding practical knowledge.

dogmatic or systematic decisions in metaphysical questions hinder in the least the fact that experiencing retains an *existentiell* meaning.

The older psychology, proceeding from the assumption of mental elements, was on the wrong track as a consequence of its principles when it conceived higher mental achievements as mere complexes of elements. It seems to have reassured itself by postponing for the time being the investigation of the "intricacies of mental life" as a later task. The presumption was that the sharpest possible isolation of the elements, their exact description, and the determination of the laws of their occurrence and their interconnections would simultaneously bring the resolution of the later task that much nearer. This standpoint of the older psychology has in general been abandoned. A radical alteration of the problem-situation has occurred; the point of departure and the goal have, as it were, been interchanged. If the older psychology, as a consequence of its principles, sought those experiences most meager, scanty, and devoid of meaning as objects for its analyses, then only experiencing in its fullness and completeness can serve the new tasks of psychology, just as for a full knowledge of the essence of art and of artistic production only the contemplation of masterpieces can be of any assistance.

There is no cause for any concern that rigorous empirical research will be endangered or inhibited by such an attitude. Those older investigations passed over the concrete behavior and experiencing of the human being, as we encounter it in occupation and family, work and pleasure, play and labor. By what right do such investigations claim the title empirical for themselves, and by what right do they assert a priority over another mode of observation that never loses sight of the concrete object in the course of analysis? After all, it can hardly be contested that the older psychology failed almost completely in all decisive, concrete questions. It could not even settle its best posed problems of sensation and perception. Moreover, the psychology of feelings has hardly come as far as chemistry had in the laboratories of the alchemists. One need merely consider how little Külpe has achieved in his lectures on feelings. (This does not detract from the great debt owed him by the modern psychology of cognition.) In spite of the efforts of numerous energetic investigators over many years, and in spite of well equipped laboratories, the science of psychology has not succeeded in achieving noteworthy results even in one of its most important areas. This is the result of, above all, the fact that the theoretical point of departure of the element and the practical-methodological departure point of the most meaning-devoid experience possible, are quite conceivably inappropriate for the investigation of feelings. Thus there is no need for us to shrink back and hesitate when we attempt to replace the psychology of dearth, that is, that of experiences robbed of their fullness and their relational richness, with another, perhaps better. If it is not possible to erect psychology from the bottom up, then we must begin from the top. We must do this as soon as we come to the conviction that we will not automatically achieve a knowledge of the most important phenomena of mental life—earlier disposed of unjustifiably as "mere intricacies"—merely through an investigation of peripheral problems. We may speak in this way of the most important phenomena because we have seen that even perception is already co-conditioned and co-determined by these "intricacies," and that, to mention only one example, the single thought-formations, affect-processes, and volitional impulses depend on the experience of time and its transformations as a medium for experiences in general. An examination, well legitimized in this way, of the concrete experiences and modes of behavior will leave us no doubt that the claim of an *existentiell* meaning in all experiences can certainly be well established empirically. In this context the phobias acquire the significance of a natural experiment. They perform the same service for us as do all mental disturbances in general—to make us aware of the peculiar character of the undisturbed behavior. They require us to pose the question, How would the undisturbed behavior have to be constituted if pathological phenomena of this kind were to be possible?

If we call to mind the manifold situations in which phobics can be overpowered by spatial anxiety, we become aware that they do not fear the concrete danger linked to a definite place; rather they succumb to the impression of threatenedness. One need only recall that certainly not unusual case, that someone on a lookout tower cannot bring himself to step to the edge of the platform even though it is well protected by a railing. No objective danger is present, and no such danger is even inserted into subjective reality through some kind of unconscious images. The symbolic content of depth awakens anxiety and vertigo. Every man has his natural place between the poles of threatenedness and security; he cannot alter it voluntarily. If this place is altered by external circumstances and approaches the pole of threatenedness, then anxiety enters. The phobic attempts to evade the experience of threatenedness, often in a haste that actually conjures up concrete danger for the first time. He cannot remain in the situation of threatenedness; he will make great sacrifices, in some circumstances the sacrifice of life itself, in order to avoid it. The phobic individual flees the limit (or boundary); he recoils before it. The approach to it already places him in anxiety, just as the approach to the taboo arouses anxiety in primitive man. The taboo-sensibility requires the primitive to shun the vicinity of the dangerous place to keep himself at a distance from its boundary. But he cannot always succeed in this. Occasionally external circumstances lead him once again to the feared and shunned place. Since, however, he never crosses over the boundary but only reaches it and collapses there in anxiety, the anxiety attack can repeat itself with each return of the prohibited situation. While the subjectively ready individual, with every act of self-actualization, stands before a new task, the phobic encounters the same experience again and again. The confrontation with the world is never really carried further, neither through the act nor through surrender; the neurosis is a standstill. The challenge to a confrontation, however, remains in existence; flight, or looking-away, can only temporarily suspend it. Notwithstanding the subject-object-split, the human being remains part of the world and also experiences himself as such. In its decisive features the condition of the world is also that of his own personal existence. If he closes his eyes before the condition of the world he flees his own destiny, or attempts to do so.

For the temporary success of this suspension the phobic makes use of the fundamental fact that the represented (that which is represented) can manifest itself only by means of the representative. Political boundaries are at first made visible by boundary-posts. Similarly, the shunned boundary first becomes visible for the phobic through typical situations and, specifically—the case that chiefly interests us here—through situations that repeat the circumstances in which the phobic was unable to evade a shock and in which his defences were penetrated. In many cases the threat remains restricted to the one typical situation; in others it extends to those circumstances similar in appearance or kindred in meaning to the original ones.

The intertwining of the representative and the act of representation has the beneficent effect of circumscribing the threat, so that outside the specific situation or situations the life of the phobic proceeds without disturbance. It is a delimitation and localization of the destructive process to one place. One could compare this to the formation of an abscess, by which the organism defends itself against the diffuse spreading of infiltrating pus irritants. Somatic medicine utilizes this natural healing tendency by opening the abscess. Just as medicine, in doing this, is scrupulously concerned that the intervention not be the cause of a further spreading and broadening of infection, psychotherapy must also exercise great caution in treating the phobias in order not to transform the localized process into a diffuse one.

The repetition of anxiety in the typical situation naturally does not refute the significance of first-time-ness for the shocking experience. The shock is not repeated in the anxiety; rather, the anxiety is first of all a reaction to the transformations of the world that took place in the shock.

Anxiety is certainly not the sole reaction to the shocking experience, no more than the form sketched out in our first example is the sole form of the shocking experience. Were we to attempt an exhaustive presentation it would be necessary to be mindful above all of those shocks, which are not related to present or future being but rest instead on a revaluation of the personal past. In conversion-experiences the past is submerged and discarded as new orders of being and new tasks emerge, and the convert awakens, as it were, on a higher stage and begins anew. In contrast, the individual disappointed with himself turns back again and again in remorse and despair to his own past; he can no longer free himself of what has once occurred. Years spent in contentment and in comfortable satisfaction of his needs appear all at once as questionable or as ultimately lost. He recognizes that he has "lost his time." Were this merely a matter of dammed-up instinctual-energies breaking through the protective dam, a relentless or plan-less search for satisfaction would necessarily follow this breakthrough. Turning toward the past, the remorse, and the despair, however, are not thereby brought into harmony. The instinctual knows no past. Likewise, in a regression only that which has already been latent since an earlier time should again become living and current. Even psychoanalysis does not derive guilt feelings and remorse from the instinctual itself but rather from the repudiation of instinctual excitations by the Superego; here, however, remorse would have to appear precisely after the final victory of the instinctual over the Ego-ideal. Can one, however, feel remorse over a meal that one has not eaten, or over a hunger that one has suffered? One may show hate and think of vengeance toward that which compels us to hunger and places us in need, but this reaction directs itself not against our own doings but against the instigator of an affliction forced upon us and repudiated by us. Even when neglected instinctual satisfaction seems to form the object of the disappointment, for example, in the old spinster, the remorse is nevertheless remorse over the lost possibilities for sensual fulfillment in her own life. Only our individual historical form (Gestalt), created by ourselves, in its deficiencies and in its perfections, remains continuously present for us. The single deeds, actions, accomplishments, the totality of the biological and psychological processes in which the objectification was accomplished, are ultimately past. The analysis of a shocking experience of this kind also leads to the conclusion, that the *existentiell* forms its own authentic content, which the phobic struggles to pass over in living.

If, as we assumed in our first example, the phobic reactions follow the shock, then this experience must have broken violently through those barriers that the susceptible individual had erected against the invasion of the *existentiell*.[60] How is this possible? After the preceding analysis of the structure and content of the shocking experience, we can well understand that such experiences, once they have taken place, determine all further experiencing. We must still clarify, however, what nevertheless enables or impels the susceptible individual who barricades himself against shocks to realize such experiences. Certainly the deficiency in objective readiness severely limits the individual's vision for the *existentiell* essences (*Wesenheiten*) represented in the phenomena or even obscures it completely. It remains to be investigated whether or not a different meaning is ascribed to the objective process in those cases in which the shock becomes a psychic trauma than is ascribed in a case of the subjectively ready individual. At this point, however, we enter the much disputed domain of questions concerning the causal or adequate connection among event, experience, and experience-sequence.

60. In the legend of the Buddha it is told that the sight of *one* beggar, of *one* sick man, and of *one* dead man destined the young prince to his holy life. Until then he had been kept distant from all knowledge of need and suffering. It would have required only an insignificant revision in our example to allow the youth to go forth from the shock grown up and matured. Of the many possible forms of the shocking experience, I have chosen the one described here because it best corresponds to the conditions on hand in an accident. The objection that pathological experiences are actually the richer ones does not, after the foregoing discussion and those that will follow, seem to be justified.

5 / *Event and Experience*

Whenever we speak of causation in the domain of natural events we mean that cause and effect are linked to the now-here-so of the things. If an individual is run over by an automobile and suffers severe injuries or dies, this unfortunate result depends on the fact that the victim in a certain place and at a definite time has collided with an auto traveling at a certain speed. Had he not stood there just where the car drove or had he not been there exactly at that moment just as the auto drove by, or had it been a doll carriage instead of an auto the result would have been an entirely different one. That is self-evident. But this self-evidence must be emphasized most vigorously from the outset in order to contrast the bond between event and experience with the causal connection, which has an entirely different nature.

When we portrayed the youth in our example as only a bystander to a disaster we assumed that he himself was not bodily affected, and we can also well imagine that he might have reacted in the same manner even if he had merely heard a description of such an event.[61]

If a mother learns of the death of her child, her sadness comes about not because she learns this in a specific moment and at a definite place, in a specific language and mode of expression, cautiously and considerately or suddenly and brutally, but rather because she understands from the communication the sense of the death of her child.

Since the incident does not have its effect as a natural-event but rather through its meaningful content, it also lacks the *Reactio* existing in the connection between cause and effect. If an auto drives into a tree the collision will damage not only the tree but also the car itself. On the

61. Cf. Jossmann's report in volume 13 of the series of papers of the *Reichsarbeitsblatt* on "The Accident or War Neurosis."

other hand the natural event undergoes no alteration in itself when someone derives a certain sense from it.

If an incident has an effect on experiencing only because a specific sense is derived (or gathered) from it, and if this "sense derivation" (*Sinnentnahme*) is independent of the special conditions of the incident, especially of all quantitative determinations, then the prospects of discovering some kind of causal, analogous connection between the event and the flow of the experience appear to be quite poor. We have emphasized this difficulty from the beginning in our example by the difference between the doctor's behavior and that of the youth.

The problem of sense-derivation is also important for the formulation of legal questions. The obligation to pay compensation can exist only when an injury has resulted because of the peculiar individual character of the injuring process. This process is the cause of the injury and has had an injurious effect only because it has happened precisely in this way and not otherwise. In the swarming traffic of the big city traffic accidents are avoided daily by only a "hair's breadth." This image graphically expresses the fact that actually the final details decide whether or not a mishap or injury will take place. Therefore the most exact possible reconstruction of a singular case serves as the basis for judgment in criminal law as well as in civil law. Everything depends on establishing exactly how an incident has factually taken place.

Even the formulation proposed by Seelert[62]— that the accident-neurosis is merely the consequence of the reaction of psychopathological individuals to the accident-experience—sharply emphasizes the independent nature of sense-derivation, at least in certain experiences. In the expression "reaction," the accent certainly lies on "action," on the autonomous, active assumption of an attitude by the individual. In any case, the entire problem of the connection between event and experience is once again obscured because Seelert's presentation relates the independence of the sense-derivation to the psychopathic character of the accident-neurotic individual. If, however, there is a general human possibility of actively assuming an attitude, this cannot be derived from the psychopathology; the psychopathology explains only the individual character of the attitude assumed. Seelert's expression, "a reaction to the accident-experience," leaves in question

62. *Deutsche medizinische Wochenschrifte* Nr. 10, (1927).

whether or not he wishes to restrict the independence of the attitude to the further development of the original experience. In any case, an entire range of clinical and psychological researchers would generally contest the independence of sense-derivation for the original experience as well as for perception; they would assume instead that perception stands in a simple, direct dependence on the stimulus.

Nevertheless the clinic assigns to the fright reactions a special position among the sequellae of accidents. Surely it does this not solely because after great catastrophies fright reactions have been found in almost all touched by them, thus refuting the objection that these are peculiar reactions conditioned by a psychopathological constitution. Even the symptoms most extant in such cases—alterations in the body, particularly in the vasomotorium—are not the ultimate basis for the attitude of the clinic. On the contrary, first and foremost, the observations cited, supplemented by the psychological experience of everyday life, have led to the conclusion that there is in the reactions to catastrophes a *coercive connection* between the natural process and the experience. Nevertheless, in relation to the totality of the psychogenic reactions—which Bonhoeffer[63] has described and differentiated from hysterical reactions—the following remains to be considered. The mental reactions following a severe fright—the stupor, the delirium, as well as the alteration of the vegetative functions—occur when a certain subjective reality has formed in the experience. This connection between transformations of subjective reality and alterations of the mental and physical functions belongs to the problem-domain of psychophysical coordination, not to that of psychogenesis. We enter the problem sphere of psychogenesis when we ask, How is the subjective reality constituted, and what role does the external process play in this constitution? The clinic's answer regarding the fright reactions is that the natural process is decisively important in this area, that is, there is a coercive connection between event and experience.

We can concede this without hesitation; we can even go one step further and grant that such a coercive connection is comparable to the causal relationship, although not identical with it.

Thus on the basis of clinical experiences we must evidently limit our claim that sense-derivation is independent of the now-here-so-givenness of the external process, and must concede that the sense-

63. Bonhoeffer, *Allg. Z. Psychiatr*, 68 (1911).

derivation is only relatively independent. One specific sense cannot be derived from any arbitrary incident, and any arbitrary sense cannot be obtained from one event in its specific, present particularity.[64]

Thus we now concede that sense-derivation is only relatively independent or, to say the same thing differently, is relatively dependent, so that we may speak even of a compulsion, based in the event, toward sense-derivation. Nevertheless we hold firmly to the fact that the original experience and even the simple perception are in the same way equally free and bound.

There is extensive literature concerning the relationship of perception to sensation. I cannot go into these lengthy discussions; instead, in order to establish my point I wish to mention only one factor that seems especially important to me, namely, the presence of deceptions in the entire region of perception.

In causal relationships there are no deceptions. Certainly we can deceive ourselves as to whether a causal relationship exists between two processes. But in the ontological sense it is not possible to inquire after the existence of deceptions in events standing in a causal relation. The presence of sense-deceptions, illusions, and hallucinations shows that what manifests itself in perception does not have an effect causally. On the contrary, the experience is determined by its sense or meaning-content and indeed, in the case of perception, by the insertion of an appearance into the region of nature. Perception, through its quality as perception, guarantees that an appearance belongs to the order of nature. More correctly stated, it appears to guarantee this.

If a fire were to occur in a theater performance in the course of the action, and if the fire were thus part of the play, the cry of the actor "Fire" or "It's burning," would not drive the audience into wild flight out of the house. The panic, which would break out if the same cry "Fire" were to ring out in the audience, would be absent. Both are perceived, but in different spheres of being: the one as belonging to a play, to a mere phantasy-formation; the other, to reality. Children in the theater easily forget the character of the play, and it is almost a regular occurrence that children call out and warn the hero who is in some kind of danger or seek in some other manner to intervene in the play, just as if it were reality.

---

64. Pathology shows in the "ideas of reference" how far the limits of independence are actually drawn.

Thus, in the context of the problem of perception, we can discuss the questions, In which cases does a compulsion exist toward the derivation of a specific sense, What structure do these cases exhibit, and Do the accident-incidents belong to this group? Raising these questions will enable us in the following presentation to again take up the ideas that we applied in the third chapter toward a critique of the genetic theories, and to develop these ideas further in a positive fashion.

If we conceive perception itself as the primary act of sense-derivation, we order it into the system of time immanent in experience and interpret it as a transformation in the course of individual history. On the other hand, if we maintain that there is in many cases a compulsion toward sense-derivation, we bring perception into the greatest possible dependence on the details, on the now-here-so of external processes. In this case the entry of a transformation in the region of individual history is determined by the objective event, which includes even the impression things make on our sense organs. We can accordingly narrow our task to the investigation of the relationships between perceptions—as transformations in experience-immanental time—[65] and the events occurring in objective time; that is, we will investigate the dependence of perception on objective events. We will attempt to resolve these questions with the help of examples that are the simplest possible, yet close to life.

If there is supposed to be a relationship between an event and an experience, which we call compulsion toward sense-derivation, then perception must already be compelled. How, however, can a process compel its being-perceived? Does it not seem that a compulsion could only commence in the first place to the extent that something is already given? I nevertheless am explicitly of the opinion, that the compulsion to be-perceived does not differ in principle from the deeper sense-derivation.

In order to elucidate this opinion, we wish to carry the image of the theater fire yet a bit further. We wish to assume that in the course of a performance flames leap up somewhere on the stage but that the action of the play does not define this fire as part of the drama. The most probable result is that at such a sight the audience will be seized by panic and try to reach the exits. Thus in this case the occurrence of the

65. Cf. Straus, *Das Zeiterlebnis in der endogenen Depression und der psychopathischen Verstimmung* ("The Experience of Time in Endogenous Depression and in Psychopathological Mood"). *Mschr. Psychiatr.* 68 (1928).

fire on the stage compels a definite mode of experiencing and behaving, similar for the entire audience. From the harmony of the behavior we may first of all infer empirically-inductively the compellingness of the impression. However, several more stages of the sense-derivation still remain to be differentiated. The flight from the theater is the reaction to the experience of threat, a sense or meaning that perception has in turn derived from the flames. But the harmony in the audience's experiencing extends still further. During boring points in the drama one person will overlook this aspect, another, that aspect of the dialogue and events on the stage, but the flames are perceived by all. The harmony is thus a threefold one. The event of the fire, the natural process, compels a harmony of perception—the primary sense of the fire; a harmony of the deeper sense-derivation—danger to life; and a harmony of the reaction—flight. But why are the flames noticed by all, while some banal process in the plot eludes a part of the audience?[66] The answer to this question is that even before the flames are perceived as flames they announce themselves to all as an incident in the perceptual sphere, or the sphere of reality.

The human being is continuously turned, inquiringly or expectantly, toward the sphere of perception. It seems possible, of course, to fully forget the environment around one while engaged in a captivating conversation or to sink entirely into reflection. Any kind of alteration in the environment, however, will break up this captivation all at once. The perceptual world again steps into the foreground of experiencing. The experiencing individual, guided by the event, turns toward it with new questions. The objective alteration, however, would not be experienced as event unless, even before its entry and in spite of the condition of absorption, the environment had already been comprehended along with it. Thus, as long as nothing new objectively happens in the environment, the inquiring expectation remains unanswered or, more correctly stated, the environment gives a negative answer of "being in order." Therefore the alteration giving rise to renewed perception, measured physically, can be entirely trifling. It is not necessary that the sense of the event be founded by a process of greater physiological intensity. Yes, it is even possible that the cessa-

66. Our example would perhaps seem still more faithful to reality if we were to assume that many are first made aware of the flames by the anguished cries of the others. The generality of the compulsion proceeding from the incident is not narrowed by this new assumption, as will soon be made clear.

tion of a physiological stimulus signifies the event. Event here means alteration of the environment, and this can also be designated by a diminishment of objective processes. If the engine noise of an airplane is suddenly interrupted, this stillness will be perceived as announcing trouble, and the question will arise: What has happened? During acrobatic performances in a variety show, the musical accompaniment is often interrupted before the high point of the performance in order to heighten the tension of the audience by the suddenly commencing stillness. This stillness is felt as torture. Inversely, when the sphere of hearing is filled constantly, such as by the splashing and murmuring of a brook, this has the effect of soothing and lulling one to sleep. The effect, however, is certainly not a purely physiological one; rather it results from the sense of an environment present to us in sonorous fulfillment and free of threat. If we were to close our eyes in order to abandon ourselves to sleep, the communication with the environment would be sustained by the noises and sounds that reached us. We do not wish to relinquish communication with the surrounding world even in falling asleep and in sleep. Complete stillness, the "deathly stillness," interrupts this communication; it is precisely by this means that it can disquiet one and disturb sleep. The constant murmuring of the brook, however, soothes us and lulls us to sleep, not because it tires our ear, but rather because it *says* something to us—promises us that "Earth, this night too thou art abiding" (Goethe, *Faust II*). The constant buzzing of flies, on the other hand, becomes an annoying disturbance of sleep, even though the buzzing, again defined physically, can be considerably less intensive than the murmur of the brook.

Perhaps it is permissible to cite yet a few more examples. A bang that suddenly breaks the stillness arouses fright; if the bang were to be repeated at not too lengthy temporal intervals and nothing were altered in any way in the physical and physiological conditions, the original effect would fail to occur. Although even in the repetition the bang contrasts sharply against the stillness, and thus is sudden in the objective sense, it is nevertheless experienced as belonging to the accustomed environment, and the experience of the Sudden is missing. Once again, here we see what we have already referred to above, that the Sudden is a definite content of the experiencing and is relatively independent of the modality, quality, and intensity of the stimulus. There are conceivable circumstances in which a sudden bang, that is, the starting shot, will not excite, but rather it will equalize tension for the two participants in a race. The "stimulus" certainly does not have its effect as such but rather because it carries the meaning "event," and it does that not as a single stimulus but only in *relation to the preceding stimuli*.

Since the individual stimulus accrues its meaning only from its relation to the preceding ones, there is an interchangeability of stimuli. The neutral (indifferent) and the nonneutral (different) states can be built up from the same elements. Suddenly commencing noise and suddenly commencing stillness, according to the preceding circumstances, can both call forth the same mental effect. There is an uncanny and a comfortable, a pleasant and a distressing, stillness. Should the noise follow after the stillness the noise would carry the meaning, "Attention, something is happening here"; inversely, should the stillness follow after the noise, the stillness would carry the same meaning. The contrast could not be experienced, however, if the meaning-poor situation, which precedes the meaning-ful one, had not also been given, that is, if it had not also been experienced.

Perception thus does not begin in one definite point of objective time in order to end in another and to remain at a null level until a later reappearance; rather it *fluctuates between neutrality and nonneutrality, indifference and difference*. The perceptual world sinks down into the mode of givenness of neutrality only to come into prominence again with the entrance of the contrasting stimulus, as differentiated and to be differentiated. Throughout these fluctuations the perceptual world remains abidingly *placed in question*. In the neutral interval the question receives a soothing answer, which makes a more deeply penetrating questioning superfluous. Only when the contrast produces a sudden change into a nonneutral condition, does further questioning become necessary. Such more deeply penetrating questioning then throws the contrasting stimulus—and everything that appears in connection with it up to the point of the reappearance of neutrality—into sharp relief against the usual context. The questions are followed now by answers divergent and distinct from the usual. We identify this process of penetrating questioning and of distinctive answering by the terms "attending" and "observing." As these terms suggest, psychology has customarily directed its attention almost exclusively to the answers. The questioning behavior and the dialectic nature of the whole process of perception, however, have been overlooked by psychology. As a result, psychology arrived at the erroneous conviction that nothing at all is given in the neutral condition. It overlooked the positive content of the neutral condition, the peculiar character and structure of the questions and answers present in the neutrality. The active moment in sensibility is thus too little, or to be more exact, not at all considered in the physiology and pathology of the senses, even though many disturbances—I call to mind here only the hysterical disturbances of sensibility—can be explained by reference to the dialectical process of questioning and answering. Further, the sensations from the sense organs can be analyzed into this dialectic process.

Within certain limits all stimuli of any kind, regardless of modality and quality, can enter into neutrality. Habituation to the regular presence of certain stimuli, which allows them to subside into neutrality, does not rest, however, on a physiological blunting or fatigue of the sense-organ or of the brain. It rests on an insertion into the sensory-region of neutrality, that is, of what is currently meaningless. Fatigue plays no decisive role here, as is shown clearly by the fact that the everyday stimuli of our environment remain neutral even when we find ourselves fresh and well-rested in our usual surroundings. By this statement we naturally do not deny that in general there are processes of fatigue and that they elevate the stimulus-threshold, but we also know how meaningful occurrences can banish fatigue and even sober up someone who is intoxicated. To banish fatigue or to sober, however, signifies precisely that meaningful events compel one to renewed questioning and to sense-derivation. This effect is nevertheless possible only because perceiving stands continuously under the sensory relationships, which in such moments stand out all the more pregnantly; even dangerous situations can become neutral for us with persistent, lingering presence, as is shown daily by the thoughtless, frivolous behavior of workers in dangerous industries. Duration and repetition *found* the indifference, that is, the sense of meaninglessness; they do not *cause* it. The limit within which habituation can result depends on the extent to which the existing stimuli restrict or do not restrict the individual's natural freedom of movement in the broadest sense. The noise of the motors in the airplane can lull one to sleep but remains disturbing for someone awake; it does not become neutral for him because it makes conversation impossible for him. Beyond a certain limit of intensity the stimulus remains disturbing. Duration and repetition do not in this case lead to neutrality because here the stimulus hinders the actually desired activity. Thus the precise goal of the activity chosen determines whether or not certain stimuli, to which a habituation already exists, again become disturbing and emerge from neutrality. I presume that it is possible to determine experimentally and quantitatively where the boundary runs between neutrality and disturbance, with regard to natural modes of behavior and to lingering or temporary tasks.[67]

Smells that fill a room usually cease to be perceptible after they have lingered only a short while. The moment of attraction or repulsion is characteristic of smell to a particular degree. Should one not yield to this tendency the possibility for discrimination would also disappear forthwith. Here too factors of meaning are in play along with physiological fatigue. We meet entirely similar phenomena involving noises. Presumably the same relationship would be demonstrable in the optic sphere if one were to succeed in filling a room as homogeneously with colors as it can be filled with smells and sounds.

A wild hunger can become the dominating object of experiencing. But here too the organ-sensations do not work directly on the thinking or the cerebral processes; it is the meaning sphere of embodied existence that determines the experiencing through the occurrence of hunger. Thinking does not simply cease under such conditions; it circles around the object of nourishment or directs itself toward the possibilities of its procurement. If we leave aside the extreme cases of hunger-need and the immediate threat to existence, hunger will not necessarily have such an effect. Submerged in work or gripped by expectation one can fully forget hunger. When this happens the organ-sensations have not simply stopped; nor have they altered their quality. If one's experiencing is no longer captivated in some other way or if one's anxious expectation is brought to a conclusion, the long-present hunger will be noticed all at once as a stormy craving. Thus even the organ-sensations can be neutral in perception. The diversion of the

67. An impairment of the capacity for certain functions (for example, the hard of hearing or the bodily afflicted) will constrict the circle of neutrality. What was neutral will emerge from neutrality once again and will be experienced with ill-temper, annoyance, irritation, mistrust, and paranoia. The hard of hearing individual who sees others speaking but cannot himself understand cannot deal with the conversation in the same manner as someone in full possession of his faculties. Someone with full hearing can dispatch the whole conversation into the sphere of neutrality because he could hear it if he so desired. While someone with full hearing can ignore the conversation, the hard of hearing must always notice to see whether or not speaking is going on. The hard of hearing individual struggles for a lost connection to the community, while the blind man is condemned to fight his surroundings for the preservation of a narrow region of private existence.

organ-sensations, however, is possible only because experiencing is always determined by sense or meaning and because the impressions mediated by the sense organs or common feeling (*Gemeingefühl*) constitute only one sphere of being—precisely that of natural existence. If the surrounding world is determining for experiencing, then inner becoming occurs together with the external process. The individual historical transformations follow the course of external incidents. The Now of transitive time determines the passage in experience-immanental time.[68]

The organization of what is given into neutral and nonneutral continues in the region of meaning. Something neutral or indifferent is something with no current meaning for my becoming or something settled, either fully settled or temporarily suspended, insofar as the sense of becoming determines what will be present and of current interest for me. Only that which can be fit into the system of meaning can become neutral or nonneutral. Sensations can become neutral only to the extent that they are vehicles of meanings—of nature, of the human environment, of embodied existence. The organization of the given into neutral and nonneutral cannot be explained by any physiological theories of inhibition, pathways, or exhaustion. The hunger example has taught us something very important in this regard. Namely it shows that the same principles worked out above for the sensations from the sense organs are also valid for the organ-sensations. Hunger does not tire, and the organ-sensations do not cease, no more than do the physical impressions proceeding from the objects of our environment cease to impress themselves on us. The organization into neutral and nonneutral is thus not the result of any kind of causal process such as the phenomenon of fatigue.

Furthermore, fatigue and neutrality can be readily distinguished in the experience. For the experiencing individual the neutral objects are settled and done with. They no longer have the character of the summons or the "worthy of question." The situation is entirely different with fatigue. While at the beginning of a journey in an especially beautiful landscape such as in the Dolomites the participants incessantly call each others' attention to the beauties of the region and the conversations do not break off, after a few hours an ever-

---

68. Straus takes this contrast between the transitive time (*transeunte Zeit*) of the world and the immanent time (*immanente Zeit*) of my inner experience from the cognitive psychology of Hönigswald, as Minkowski pointed out in his *Lived Time*. (Translator's note.)

growing tiredness comes over the entire company. The exclamations and the pointing stop more and more; simultaneously one traces the fatigue, and one feels no longer capable of being receptive, that is, no longer capable of fulfilling the demands proceeding from the object. Thus there is in principle a difference between someone who is fatigued by long work at the microscope and no longer able to distinguish the details of a preparation exactly, and a psychiatrist who is accustomed to the restlessness of the mentally ill and will not allow himself to be disturbed by a restless ward in his observation of a patient.

Neutrality is nevertheless not limited to receptive behavior. Even our movements can, as the so-called automatized movements show, be neutral. Not only rest but also standing and going can sink into indifference. For many it is a condition of their productivity that while absorbed in reflection they can go back and forth or wander. As they do this the environment becomes just as neutral for them as it is for anyone accustomed to reflect while sitting or lying.

The question concerning the connection between event and experience, as we can see, touches closely upon the question concerning the conditions for involuntary attention. Much of what we have reported here belongs to the subject matter of the psychology of attention. It is not presented there in the same perspective given here, but then the doctrine of attention, particularly in its theoretical portion, belongs generally to the most dubious chapters of psychology. The difficulties making it impossible for the theory to unify the contradictory phenomena under one viewpoint come about mainly because the majority of researchers, even though they no longer hold firmly to the constancy principle, nevertheless view consciousness as a coalescence of many single processes succeeding one another in objective time. Characteristic for this attitude is the empiricistic theory of temporal signs, as advanced by Lipps. Even the mode of expression of act-psychology,[69] which speaks of single acts, promotes the erroneous belief that psychology deals only with an objective succession of

---

69. *Act-psychology*: Franz Brentano (1838–1917), a German psychologist, founded his Aristotelian psychology on the *intentional* structure of the mental *act*. The mental act refers to, or *intends*, an object outside itself, that is, it is intentional. Brentano's act-psychology, presented in his *Psychology from the Empirical Standpoint* (1874), is the principal forerunner of Husserl's phenomenology. Both Brentano and Husserl aim continuously to maintain the distinction between the mental or psychic activity (the subject side) and the field of objects that it intends. (Translator's note.)

individual mental processes. Also in accord with this is the formulation in Froebes' textbook, which synthesizes the various views explaining attention.[70] Froebes believes that the most satisfying answer to the question as to which object is denoted by the words "attentive, un-attentive, and scattered" is the answer that these words concern the prevailing activity in present consciousness.

Things look very different, however, as soon as we comprehend that our concern is not with present consciousness but with the conscious-ness of the present. In other words, the consciousness of the individual person unfolds as the experience of his own inner history. Every single moment is a phase in his historical becoming. Everything coming into consciousness in a specific moment is determined by how it fits into the course of this becoming or how it arrests or runs contrary to it. Everything attention lays hold of, is present and is now. But this Now is the Now of the inner life-history, whose progress in becoming is not measurable by the standard of objective time.

Let us take a simple example: A schoolboy who does not follow the lesson, who allows his thoughts to wander here and there, and who "dozes," according to the terminology of the theory of attention, is inattentive, scattered, and a *"distrait-dissipé"* (distracted-dissipated type) in the sense of Ribot. The concentration of his attention is little; the distribution, great; and the constancy approaches the null-point. This statement, however, has still only incompletely comprehended the essential nature of his mode of experience. Not only is he not present, in the sense that he takes no notice of the occurrences of his surroundings, but also he is without a present; his inner becoming has halted, as in someone sleeping. If he is startled out of his day-dreaming by a shout from his teacher, then the beginning and end of his dream state close immediately together, just as in awakening we carry on at that same point in the course of our life at which we stopped in falling asleep. Only through this reference to inner becoming do the two stretches of waking existence, interrupted by sleep, join one to another without gaps.

The behavior of the distracted schoolboy, however, is absolutely different from that of the *"distrait-absorbe"* (distracted-absorbed type), that is, the inattentive individual who, thinking through ques-tions that to him are essential, likewise notices nothing of the occur-

70. Froebes.

rences in his environment. Such an individual, carried away in his own world of thoughts, is not without a present; on the contrary, if he succeeds in any moment in solving his problem, a significant advance will come to pass in his becoming. If someone is occupied in thought with some kind of historical incident, then that incident under consideration in his thinking is not present but has its place in the historical past. He as thinker, or the real accomplishment of his thinking, however, have their place in the present, in the objective time to which he belongs in reality, and which belongs or can belong to his experiencing. The experiencing is of course constantly related to reality, but is not determined by the occurrences in it. The real processes and occurrences only become determinant when they have become meaningful and of current interest.

Even in a state of neutrality experiencing always remains related to the natural environment. The environment is not annulled in neutrality; only its meaningfulness for the current experiencing dies out. The "Neutral" loses its demand character so that the experiencing can, in its content, organize itself entirely free from the neutral environment.

Through the organization of the given into the neutral and nonneutral, everything given in experiencing is ordered according to the course of becoming in experience-immanental time. But it is precisely in this way that experiencing at the same time becomes imprisoned by the course of things in transitive time.

With the sudden change of a neutral state into a nonneutral one, the external stimulus—in its quality and its quantity, from its temporal and spatial position, and thus in its now-here-so-ness—forces experiencing toward sense-derivation. Although this effect only accrues to the stimulus because of its relation to the preceding neutral stimuli, nevertheless the stimulus, in its own individual particularities, now determines the experiencing. The paradox apparently dissolves here if one keeps in mind that which we have previously stated, namely, that the organization into neutral and nonneutral introduces order into what is given, order determined temporally and in its contents by the experience. The neutral givens are not capable of determining the experience in its contents; nevertheless, insofar as the neutrality allows the experiencer the freedom to turn to something else, as an empty place it remains a virtual object of experiencing in general. With the entry of the nonneutral stimulus the experiencing follows, in its content, the events of the environment. It has already previously been related to these events but in the peculiar form of neutrality, in which the experience has the intentional character of ''being-free of something.'' The nonneutral stimulus thus does not create for the first time a bond between incident and experience; rather it alters the modality of this relationship in such a way that its ''being-free of something'' is replaced by a ''being-determined by something.'' The becoming conscious depends not on the intensity of the stimuli but on the meanings constituted in them.

We have endeavored to give, in broad outline, a solution to the problem we posed for ourselves in the beginning of this section. Our question was, How could an external process or occurrence determine experiencing, that is, compel the derivation of a certain sense? In answer, we found the decisive factor in the organization of the given into neutral and nonneutral, an organization arising from the experiencing itself. Whenever such a compulsion toward sense-derivation exists there is a connection between event and experience, which can be regarded as analogous to the causal relationship. This explains that even the unready individual can be pressed into a shocking experience, which continues its effects as psychic trauma.

By demonstrating in general the possibility of a compulsion toward sense-derivation, at the same time we establish that there must be various forms of connection between event and experience. Accordingly we have yet to examine which criteria allow us to decide, in the individual case, what kind of bond is present and particularly whether there has been a compulsion to sense-derivation. The temporal sequence alone cannot be the criterion; it is no criterion for whether two incidents stand in a causal connection. (We certainly do not consider the movement of the hand in the railway station clock as the cause of a train's departure). So too the temporal sequence of event and experience does not decide whether or not there has been a compulsion toward sense-derivation.

On the other hand, in discussing the accident neurosis, Hoche maintains that before the accident the victim was healthy; afterwards he was ill, and consequently the accident is a condition of his being-ill. This argument lacks all power of proof. The state of affairs took on such a simple shape for him because he failed to distinguish different forms of bonding between event and experience, so that for him it was effectively superfluous to seek any kind of further criteria apart from the temporal sequence. But, as we have just seen, even in the region of

physical causality certain further factors in addition to temporal sequence must be present if the coincidence of two incidents is supposed
to be viewed as a causal connection. In the bond between "eventing"
and experiencing, essential differences stand out in each of the following cases.

If a gathering for intellectual work were interrupted by the noise of
voices in the immediate vicinity, it would appear as though the noise
had directly and actively (and it is precisely this that makes this bond
comparable to the causal connection) influenced the experiencing. We
know from our preceding discussion that we are dealing only with the
semblance of a direct effect, although to be sure, if the expression be
permitted, with a semblance founded in the structure of the experiencing. Things are entirely different, however, if while regarding the
figures on a chess board in the position just arrived at, someone would
find a beautiful combination for the further conduct of the white or
black player. Plainly there is no longer any compulsion toward sense-
derivation here. Such a compulsion is present only when alterations
take place in a medium in which we continuously participate—a
medium, such as nature, in which our own existence actualizes itself.

When an event exercises a compulsion toward sense-derivation, the
sense, whose derivation is compelled, and the direction of its further
unfolding must also be determined in their contents. The compulsion
toward sense-derivation signifies that not just any arbitrary kind of
sense can be derived. In order to discriminate among the different
forms of sense-derivation, I wish to introduce here the expressions
ascending and *collateral* sense relationships.

Our basic example shows a case of the ascending sense-connection
as ascent from the sight of a dead man to the universal meaning of the
existence of death in general. On the other hand, the famous anecdote
connecting Newton's discovery of the laws of gravity to his experience
of a falling apple offers an excellent illustration of the second kind of
connection—that of the collateral sense-derivation. The examples are
pregnant enough to display the essential distinction without an exhaustive discussion of the many problems involved here. They show that it
is not legitimate to claim every experience as a causal consequence of
an event merely because it refers, in its content or through its temporal
position, back to this earlier experience. If Hoche were right, then one
would have to say that before the fall of the apple Newton had no
solution to the problem of gravity whereas afterward he had it, and

consequently the fall of the apple is a condition for its discovery. If the Newton anecdote were true, however, it would show precisely how productive experiencing seizes upon an incident and does not allow itself to be compelled by it to a specific sense-derivation, but rather approaches it actively and inquiringly, drives a neutral process out of neutrality, and again discovers the problematic of everyday life. It is precisely in this way that the particular case is comprehended from the outset, not as this single, natural event, occurring only once and bound in its individuality to its spatial and temporal position, but as the *representative* case. Not until Newton discovered the problematic contained and represented in it was the experience lifted out of its everydayness.

In the theater fire, on the other hand, it is actually the one-time occurrence, the being present and becoming affected by it, that are decisive for the sense-derivation and the configuration of the experience. Here too a sudden change occurs from neutrality into nonneutrality, which the effected individual cannot escape. What is at stake after all is the sphere of the natural foundation, the sphere of existence, toward which we are continuously, inquiringly, and concernedly turned. The fright, which effects the change from neutrality to non-neutrality, compels the ascending sense-derivation. The individual affected is unequivocally forced by the process, in its first-time-ness and in its now-here-so-ness, into the shock and into further experiencing.

6 / *Neurosis and Self-Actualization*

Thus far we have analyzed the shocking experience and—in our last chapter—discussed the compulsion toward sense-derivation as well as the ascending and collateral sense-relationships. This puts us in a position to consider the concrete experience-configurations and modes of behavior in the individual case, and to ask what kind of bond exists between them and certain events of the past to which they refer. We wish to make use of this possibility by turning back to our starting point in order to apply these newly won criteria to the compensation-neurosis. We wish to ascertain whether it is legitimate in the strict sense or in a mere analogy to characterize the accident-incident as the *cause* of the neurosis.

The result of this investigation will, we wish to anticipate, be essentially a negative one. It will show that we can speak of a compulsion toward sense-derivation, which we must put in place of the causal connection, only in a narrowly limited number of cases. We cannot suffice ourselves, however, with this negative result. Those who explain the compensation-neurosis as an effect of the accident did not arrive at their conception through an analysis of the relations between event and experience but inferred indirectly the causal significance of the accident. They sought to prove that the interpretation of the compensation-neurosis as an expedient or purposeful neurosis and the allusion to the images of desire and to the constitution are inadequate to explain its occurrence. The mere rejection of a compelling bond between accident and neurosis would thus at this point not carry the discussion any further. We will therefore attempt to expand and deepen the prevailing theory. To this end, however, we must backtrack considerably one more time at the end of this chapter and submit the deforming tendency, as a basic feature of neurotic, perverse, and addicted behavior, to special examination. Only in connection with this will we be able to bring the discussions of the compensation-neurosis and the psychic trauma to a conclusion.

The possibility of directly examining the nature of the connection existing between an event and an experience makes it easier for us to take up a position on the very controversial question, What significance has the constitution, especially the psychopathological constitution, in the formation of the compensation-neurosis? The retention of the concept of constitution in the form put forth by Martius has created a series of apparent problems. Martius formed his concept of constitution in order to take certain experiences into account with regard to the pathogenesis of illnesses. The impetus for his theory came above all from the observation that many of the external causes of illness do not by any means cause illness in all individuals who come into contact with them. There must therefore be additional individual peculiarities for an illness to break out. For Martius, the concept of constitution is incorrectly narrowed down to these individual differences. One thereby gets an impression that illnesses, like autonomous things, were produced by the collaboration of external and internal causes.

As I have already emphasized,[71] however, we must in principle assume as many different varieties of specific illnesses as we delineate constitutional types. Each type has its illnesses. For the physician it is purely a practically based artifice, that in spite of the distribution of individuals treated by him into single varieties and types he stays with a fixed generic concept of illnesses and seeks after individual factors for the generation of these illnesses. This artifice makes possible the formation of a fixed schemata for prophylaxis, diagnosis, prognosis, and therapy. Nevertheless this procedure is theoretically false. We can no more call the constitution the cause of a disease than we can call iron the cause of rusts, or copper the cause of verdigris. Just as the rust and the verdigris arises through atmospheric influences, so too illness processes occur in single individuals through the action of specific factors. The constitution determines those factors that are in general able to have an effect and those manifestations that they will call forth.

Precisely the same principle is valid for the relationships between accident and constitution. The incidence of accident-induced injury in

71. Cf. E. Straus, *Das Problem der Individualitat: Die Biologie der Person*, Bd. I (Berlin-Wien: 1926).

each case depends, on the one hand, on the constitutional condition of the organism, and on the other hand, on the occurrence, which is quantitatively measurable as a natural event. The accident always signifies a displacement of the boundary from the environment toward us. The sudden breaking through of the boundary (limit) of adaptation produces an injury of the organism. Thus, to cite only one example, if the heat in an electrically driven furnace were suddenly to rise as a result of a bad connection or a defect in the circuit, the workers occupied in this workshop could suffer burns or other heat-induced injuries. The increasing temperature in the room exceeds the physiological range of adaptation, whose most extreme limit, always established by the constitution, has been defined by a rather exactly definable position on the thermal scale. We must, however, distinguish further between the typical environment and the individual environment of a human being. In these matters confusion usually prevails between the concept of constitution, which refers to the totality of biological potencies, and Martius' concept, which refers to the individual peculiarities. Further, there is an intermingling of the juridical and medical modes of thought and observation. Even in the expert appraisal of the organic consequences of an accident, a question still arises concerning the part played by the constitution. We refer here especially to those cases in which a physical injury or death does indeed stand in a definite, causal connection with an external influence, but the dimensions of this influence do not exceed that which is otherwise generally endurable. In a man with alterations in the condition of his bones due to tabes, even a relatively insignificant cause, even a powerful handshake that another could readily endure, can bring about a bone fracture. The individual with tabes also has his range of adaptation; however in comparison to that of a healthy man it is considerably constricted. In a legal process in which liability or the question of bodily injury were to be decided, the perpetrator would have to be acquitted because he had indeed overstepped the individual limits of adaptation, which had been altered by the tabes illness, but not the typical limits. Since the injured party lives in a typically structured environment and can demand only a typical behavior from the perpetrator, the perpetrator has committed no transgression of the limits drawn by legal norms. Notwithstanding the injured party has suffered an accident, namely because, as a result of his illness, he was no longer adapted to the typical environmental limits within which he lived. In an individual private environment it would be

easier to recognize that the accident-incident sufficed as the condition
for a quantitative increase and for a breech of the limit as a result of that
increase. In each case the range of adaptation is a relation between the
constitutional biological potencies and the processes of the environment.

The administration of justice takes as the basis for its decision not the
individual limits of adaptation but the typical limits. It is only for this
reason that a certain group of accidents, in which the range of adaptation is constitutionally more narrowly limited than average, is excluded
from the right to claim compensation; at this point a standard numerical
figure, proven by experience, simultaneously enters into the quantitative determination of the accident process. In many cases of organic
injury the accident no longer forms an ''essential cause'' for the
administration of justice. Certain cases of accidents cease to involve
liability for compensation in the juridical sense, not because only in
them has the constitution become an essential factor of the accident,
but because these cases involve singular variations of the typical constitution. There are adequate causes only with respect to adequate effects. Physically speaking, every cause has its effect. The selection of
the cause thus is preceded by the selection of the effect, that is, its
definition by insertion into the biological order or into the region of
technical, social, and cultural objects. Thus when we speak of an
adequate cause, we no longer refer to a causal connection (in the sense
of physics). To illustrate this with an example, if someone were to
damage a valuable picture by splashing it with an acid the physical
process would exhaust itself in the decomposition and alteration of the
color pigments. If we were to assume that plain colors were so
irregularly distributed on a canvas that by chance exactly the same
color tones were present in the same dimensions at the same positions
on this surface as on the valuable picture, then a spraying of these
places with an acid would have exactly the same effect, physically
regarded, as the damaging of the picture. Physically no more has
happened in the one case than in the other. The process is imbued with
a special significance, however, when damage is done to the picture
merely because in this case the colors serve as the foundation for the
aesthetic object. Similarly in the illness and in the accident the adequa-
tion is determined by the peculiar nature of the organic events and of
their lawfulness.

By the same token, the construction of a factory is not accomplished only according to technical viewpoint; practical experiences with industrial hygiene must also be considered. Industrial inspections are supposed to watch over and assure that the factory is layed out in such a way that the influences on the organism of the worker do not exceed a definite limit; just as in the construction of stairs care is taken that they conform to the security and comfortable stepping of the user. If we begin with a typical boundary between environment and organism, we can then establish how much force must be applied before an event can be designated as an accident liable for compensation—just as at a customs-boundary the importing of a small number of certain goods is free and only the importation of a greater quantity becomes liable to duties. Ideally, the accident-boundaries could be exactly established, but practically the results are often only estimates.

As always in such cases the juridical decisions are inclined to omit, whether they accept or reject the claims, that the constitution remains an essential factor even in the definitely organic consequence of an accident obliging compensation. It is possible to draw limits for the bodily injury by introducing a standard figure proven in practice. For the juridical evaluation of the connection between event and experience, however, there remains only the possibility of falling back on a noncodifiable norm of behavior. This state of affairs has already produced more than a few difficulties and has brought about division among the experts and their grouping into two hostile camps. It would therefore be a significant step forward if, with the assistance of the criteria mentioned earlier, we could directly distinguish different kinds of connections between event and experience so that the question concerning the part played by constitution in an experience-configuration could once and for all be disposed of in general. No evaluation of compensation-neurotic behavior is necessary to decide the question of whether or not a compelling connection exists between event and experience. Even the objection of Hoche, in itself correct, that one may not require heroism of anyone, is no longer to the point. Only when we have answered the question concerning this connection—and it will be answered in an affirmative sense only for a small group—will the problem of a "hostility toward the norm" (*Normwidrigkeit*) surface again in the analysis.

Accordingly, our next task is to organize the totality of these cases. In the literature, *the* compensation-neurotic is often discussed. This expression is certainly too summary. Many differences of opinion originate, as the recent exchange of views in the Ministry of Labor has shown, because the interpretation oriented toward the concept of a desire for compensation does not do justice to all cases. For the overwhelming majority, to be sure, this interpretation correctly emphasizes what is juridically and medically decisive but without exhaustively answering the totality of the psychological questions that are raised in the process.

All participants in the "course" were actually in agreement in this respect. The misgivings with respect to the prevailing administration of justice extended in practice almost exclusively to the handling of a small minority of compensation-neurotics. In a theoretical respect, of course, much criticism was aimed at the medical psychology of the compensation-neurosis and at the jurisdiction, but no new viewpoints were brought to light. In an attempt to organize the total material, we can now differentiate two groups from the outset: first the phobics, who, as our exposition showed, occupy a special place; and the malingerers, who almost noone but Landauer will regard as ill. The claim that no compelling connection exists between malingering and the accident event requires no proof. The question is thus whether, aside from phobic reactions, there are still other cases in which a compelling connection exists. In classifying the totality of the remaining cases we proceed from the clinical experience that as a rule a decisive turning point commences with the settlement of the legal proceedings. The catamneses of Panse and Malling show, in spite of the doubts advanced against them, that the final settlement is the most important factor and that acknowledgment or rejection of the claim leads to the same result. If the proceeding is brought to a close both the combat-attitude and the complaints and organ-neurotic symptoms will sooner or later fade away—often only to crop up again in the case in which a pension is granted when a follow-up investigation is scheduled. The climax begins with the settlement of the proceedings. The acknowledgment of the claim in many cases makes a malingerer out of the neurotic or leads to a behavior that is at least the next closest thing to malingering. It is not legitimate, however, to conclude from this change that, prior to the settlement of the proceeding, the disturbances had also been mere malingering.

If there were only this one kind of development, that is, the disappearance of the symptoms after the acknowledgment or final rejection of the claim, then there would be no need to look beyond the desire for compensation to explain the formation of these neuroses. But there are other kinds of developments even though their number is limited. There is, for example, a continuance of the symptoms in spite of the final settlement of the proceedings. Among these cases the most remarkable group is that in which the symptoms remain in existence even after a pension is granted. Those denied a pension, even after they have been refused and even after the existing legal remedies have been exhausted, still need not finally relinquish their combat-attitude. It is highly possible that they reckon on starting up the proceedings again by some means or another and continue in their neurosis on that account. Even more striking, however, as we have said, are those who, after the award of a pension, continue their life in an often grotesque, neurotic self-mutilation. These cases do not fit readily into the explanation of the wish for a pension and lend plausibility to the presumption that additional experiences generally collaborate in the structure of the compensation-neurosis.

But which experiences? There is little hope that the pension recipients themselves could give any kind of answer to this question. Even psychotherapeutic treatment might bring only the particular psychotherapist's conception to light instead of the processes actually elapsing in the neurotic. After the experiences I have had in this area to date, I am extremely skeptical about the successful cure of the compensation-neurotic through psychotherapy.[72] In any case, such psychotherapy must take place in hermetic seclusion from the insinuations of relatives and friends. Even then we must fear that the propaganda of well-meaning and interested persons will later push them into neurosis. Recent attempts at deepening the psychology of the compensation-neurosis, particularly Von Weizäcker's interpretation of the compensation-neurosis as a "neurosis of justice," interest us only insofar as they bring into consideration the connection between the accident-event and the accident-experience. Von Weizsäcker correctly emphasizes that his own interpretation of the compensation-neurosis as a neurosis of justice (or right), as well as the psychoanalytic interpretation that he largely adopts, must lead to the denial of the adequacy of the accident-causality. Our criteria remain valid even if one wishes to assume, along with psychoanalysis, that strivings operating from the unconscious,

---

72. In the case reported in somewhat more detail by Von Weizsacker, he was also clearly dealing with a phobia. (*Nervenarzt*, 1929).

castration tendencies, or modes of behavior arising from the Oedipal situation are active in the formation of the manifest neurotic phenomena. Even when one proceeds directly from the analytic standpoint, the accident process can be fit into the continuity of experience only through a productive appropriation and elaboration.

With respect to Von Weizsäcker's presentation, it can be objected that he never sufficiently clarifies whether the neurosis develops from the desire for justice[73] or the desire for justice from the neurosis. The justice that is demanded here is making good again for a suffered injustice (*Unrechts*)—compensation for an injury. Von Weizsäcker speaks of the vain struggling against what has now, once and for all, occurred, but in many cases nothing or almost nothing has occurred. The demand for justice does not originate from an injustice that presumedly has been suffered; rather the presumption of injustice develops simultaneously with, or even for the first time in relation to, the demand. An appeal for making something good again, however, can be experienced as an appeal for justice only if the experience of the injury has preceded it. It is precisely for the subjectively honest conviction of being in the right that the connection between *appeal* and *injury* is inevitable. The demand here aims not toward the application of a paragraph of the law but at the actualization of the idea of justice, or right, in general. The stipulations created by statute law can be misused, and they are misused daily and hourly. In civil law in particular many claims are filed only because with an adept interpretation of the legal principles their success is likely. In general the claims crop up only because fulfillment beckons them. The experience of right (or justice) is lacking in such cases, or it assumes the perverse form in which misuse appears as right because only "dummies" abstain from the possible misuse of the law. The claims, however, are not motivated by the experience of injury; only then would they be justified in the true sense. On the other hand, only in the fortunate case does the claim that is brought with a firm belief in its personal justice harmonize with the stipulations of the law. The battles resulting from the collision of codified justice and personal claims to justice are familiar to the psychiatrist. The clinical picture of the querulous or paranoiac individual battling for his rights deviates from that of the compensation

73. *Rechthaben-wollen*: literally "wanting to have right." Such individuals have a strong sense of being robbed of what is rightfully or justly theirs, that is, that their rights have been violated. It is a question of right or justice. (Translator's note.)

neurotic insofar as the sham-histrionic quality prevails in the conduct of the latter. This alone should critically persuade us against imposing the honorific and therefore misleading characterization of ''neurosis of justice'' upon the compensation-neurotic. Obviously it remains conceivable that a few individuals, as a result of an influence of childhood or of the life of a proletarian existence, may for the entire duration of their lives feel themselves to be injured, deprived of their rights, and wronged and that they stand in a continuing protest against society, from which they secretly or vociferously demand and expect a making-good-again of the injustice suffered. Their claims, however, remain vague; their demands, rhetorical—until the accident gives them an opportunity to localize these claims, to concretize them, and to substantiate them with numerical exactitude. Psychoanalysis, and probably Von Weizsäcker also had these kinds of cases in mind, although in his presentation he attributed too great a significance to the accident incident itself, whereby, contrary to his own intention, he gave the impression that his ''neurosis of justice'' involves a compelling, rectilinear, and ascending sense-derivation from the event of the accident. Actually we are in agreement with Von Weizsäcker that the interpretation of the compensation-neurosis as a neurosis of justice cannot attribute to the accident event this kind of a position in the sequence of experience. The desire for justice grows not directly from the living-through of the accident, but subsequently seizes possession of the completed experience. If we limit the effect of the justice-motive to a small group of cases, the gaps in the psychology of the compensation-neurosis stand open anew.

If, in order to achieve a broader perspective, we were to compare the injuries that have resulted from the institution of the Social Security system with those of similar establishments, for example, unemployment-security, we would find similarities at many points. In saying this I leave aside entirely the misuse of these institutions and thus disregard those individuals who, only in order to be able to join in the pleasure of unemployment money, take on positions as hired laborers for a short time with the intention of being unemployed ''as soon as possible.'' Once this goal is reached, they then restrict themselves once more to their original activity, as housewife, for example—an activity they would never have interrupted if there had been no possibility of supplementary income.

Without allowing ourselves to be confused by partisan opinions, we may nevertheless probably assume that many individuals prefer a small income from unemployment relief to the higher wage contingent on work performance. The argument that is often brought forth against the interpretation of the compensation-neurosis as a purposeful-neurosis, that is, the argument emphasizing the minisculeness of the amounts paid as compensation pensions, is of no avail here. Would we view all compensation neurotics as malingerers if the disparity between the normal earnings and the pension were smaller? Von Weizsäcker says that no advantage can be obtained through the neurosis; nevertheless this is anything but self-evident. It is not true that the neurotic is not in a position to perceive "reality," that is, that "health and advantage cannot be achieved by way of his neurosis." The reality that concerns us here, however, does not present itself to perception in the way of corporeal things; rather this reality is a state of affairs that has first been created through science and law, and this state of affairs might even today gladly shape the giant and his coworkers in the direction of the neurotic. It appears dubious to me, however, whether it is legitimate to reproach all who behave in this way as work-shy or aphilopon (Schroeder's term signifying "aversion to effort"). The conception that seeks to label working in every case as working toward virtue and that reinterprets the urge to creation, which nowadays in most individuals is thwarted and neutralized precisely by work, as an urge toward occupation, renders understanding the compensation-neurosis more difficult. The situation is nevertheless unfortunately such that for most people work can be only a means to an end, and thus they must seek the authentic value of life elsewhere. Industrial work as well as the occupation of the lower salaried employees have nothing alluring in themselves. The demand for shortening the working hours is proof enough. Even if the wage-ratios were more favorable, work in its modern form would retain this negative character. A man "working at his own tasks," on the other hand, becomes indifferent to what vocation he is engaged in; the day is too short. Indeed he wishes for a lengthening of his working hours and a multiplication of his power for work. Not everyone who suffers under the meaninglessness of modern work and finally flees it, can on that account alone be disposed of as a work-shy psychopath.[74] It remains to be investigated, whether the poverty of meaning of the proletarian situation finds its continuation in

---

74. Moreover, as the modern tempo shows, working can itself become a mania.

the meaninglessness or illogic of the compensation-neurotic's desires and struggle. In any case, we may speak of such illogic only in those compensation-neurotics who, after the receipt of a small pension, regulate their entire manner of living around this pension and suffice themselves with leading a vegetative existence—with eating, drinking, sleeping, and busying themselves sexually. They retain their pseudo-demented attitude for decades, close themselves off from every value and meaning of life, and renounce every self-actualization. Such cases do appear to exist, even if they are quite rare, even rarer than one should assume from the often self-interested relatives. Compensation-neurotic behavior reaches its high point in these cases. In an earlier work I expressed myself to the effect that in such a "doubling back of the entire life line" we catch sight of a moment that is morbid and unintelligible in its deepest ground. I now believe it is possible to understand this neurotic attitude precisely as *Deformation*, and I see here the beginning of a path leading to the center of the problem of the compensation-neurosis.

7 / *The Deformation*

Deformation is a, if not *the*, fundamental concept of the psychopathology of neuroses, perversions, and addictions. A human community composed only of the perverse and the addicted could not endure; it would be doomed to destruction. It would lack the supporting foundation in the experiencing of its individual members. It would be deficient in every tendency toward the formation of a community. This conclusion is not contradicted by the fact that addicts sometimes come together in groups of their own peculiar sociological structure. Accordingly, by the concept of deformation we mean that we do not regard the self-destruction and the destruction of institutions and works as the external result of the addictions and perversions; rather we consider the self-destruction and the destruction of creations of the objective mind[75] as the secret meaning of such disorders.

It would be incorrect to assume, however, that by the concept of deformation we have merely rechristened the death instinct of the psychoanalytic terminology. Deformation belongs in the class of value-attitudes and value-actualizations; neither as object nor as goal does it resemble a drive. We wish first of all to sharpen this opposition by contrasting sadism as a variety of deformation to sadism as a partial drive, as Freud originally interpreted it. For psychoanalysis the sadistic behavior of the adult, like all perversions, is only the emergence or the revival of a partial drive; thus, just as it appears in the adult as a perversion it is also supposed to be found normally in the progress of childhood development. The sadistic behavior of adults rests on a regression to the anal-sadistic stage of the organization of libido.

75. Cf. note 20, above.

Sadism at this stage is predominantly fed by the eroticism of the muscles and expresses itself in aggression. Certainly noone will dispute the fact that at the age in which psychoanalysis locates the anal-sadistic stage of libido-development the child delights to a particular degree in the activity of his organs of movement and does not know much more to do with the objects that come into his hands than to rip them up, smash them, or break them into pieces.

Does this childish behavior in fact have something to do with sadism? Let us first of all consider so-called muscular eroticism. This is surely supposed to signify that musculature represents the organ source of a special instinct-satisfaction, a special form of gaining pleasure. The instinctual aim is realized through the activity of musculature. But if we call to mind what kind of movements these are that are experienced pleasurefully by the child as well as by the adult in sport and in dance, then we must surely be seized by doubt as to whether any kind of pleasure sensations can be brought about through musculature as such. Clearly pleasure here depends not only on the fact that the muscles are active in general but on the meanings of the movement, of the being-moved, which are founded in this activity. A half hour of uniform, continuous calisthenics does not customarily produce a condition that one might characterize as the acquisition of pleasure. Certainly the worker at his machine also activates his muscles, only neither the calisthenics nor the machine work appears in general to be reckoned among those recreations that the adult pursues in sports, and the child, in his playing with movement.

Calisthenics and machine work are muscle actions; yet there is a difference. Plainly the muscle actions and the sensations connected with them still do not by themselves suffice to explain the joy in the movement. Even if, along with Bühler, we replace muscular eroticism with the expression "functional pleasure," the misunderstanding easily arises that it is a matter here of a physiological functioning. However, the contrast between sport and work shows precisely that it certainly does not depend on the physiological processes alone but also on how the individual experiences himself in his movements, how he experiences his existence, his motility, and his relationship to the surrounding world, in particular to space. The fact that these kinds of corporeal activities everywhere and in all times give rise to an objectification of performances in sports contests and to some kind of disciplined formulation of movement shows from another perspective how much is denied by the reference to muscle activity as a source of organ pleasure.

And things stand no better with aggression. Certainly the small child takes possession of objects by destroying them. If a young puppy, however, with which we might here compare the small child, gnaws on a valuable piece of furniture or pulls apart an expensive carpet, he has certainly tried out, exercised, and activated his performance-capacity for biting and digging on these objects. We may nevertheless not seriously claim, that he has acted aggressively toward the furniture or the carpet. The furniture is for him only a piece of wood into which one can bite, and the carpet a covering surface for the ground, which is ordained for digging. The carpet and the furniture, insofar as they are valuable to the owner, do not exist for him. The aggression of the child is no different; his taking possession of things is in most cases destruction because the child still has no other form of communication with things and because things appear to him only within the order of the breakable or unbreakable as action-possibilities for his strength. Destruction is still the single form accessible to him for his effect on things; his joy at destroying is joy at his having an effect. The child as well as the human being in general experiences the objects not as disturbing stimuli, which he should like to ward off, but as objects for his formation (Gestaltung). Taking possession, aggression, and sadism are therefore also in no way equivalent. The normal erotic act of taking possession has nothing to do with sadistic aggression, even if it occasionally seems as though sadistic actions were nothing other than intensifications of actions occurring normally in sexual union. Movements that are distinguished in their course only by more or less vehemence, swiftness, and frequency can originate as actions from entirely different sentiments and can be the expressions of specific modes of behavior. Though actions in their external manifestation are distinguished only according to degree, there can nevertheless be a qualitative difference between the experiences to which they belong— in exactly the same way that there is no phenomenal continuity of transition between warmth and light, even though the physicist proves that as physical agencies the two are distinguished only by their different wavelengths. Psychoanalytic theory is voluminous here, but it overlooks completely the distinction between a value-adequate and a value-inadequate form of taking possession. One can in fact take possession of the same object in very different forms. The artist or art lover ''take possession'' of a picture entirely differently than the art dealer, the art collector, or the art historian. A vandal would perhaps use the picture as wood, if he wanted to make a fire. An ascetic zealot, however, will surrender it to the flames as the citizens of Florence did with many art works under the influence of Savanarola because in it the zealot catches sight of a work of evil and of sin. The artist, art lover, art dealer, collector, and art historian must all reunderstand the aesthetic object correctly as such; but only the artist and the art lover behave in a value-adequate manner in the face of the object. In the three other groups a value-inadequate behavior is already present; the beginnings of a deformation are at hand. Characteristic of this is the lovelessness with which the art dealer as well as the art historian often treat the works with which they are occupied; they look past the central intrinsic value of the art works. Their interests are aimed at the art work only insofar as it is either an economic commodity or an historical forma-tion. The zealot no longer lends any validity whatsoever to the aesthe-tic value for its own sake, and the vandal, last of all, doesn't perceive the picture as an aesthetic production at all but sees only canvas and wood. He comports himself like the child who breaks an object or like the puppy who gnaws it. The child and puppy and vandal destroy because the value, whose preservation is of primary importance, has by no means become clear to them. Their destruction is the product of their value-blindness. Their acts of taking possession are subjectively value-adequate; they know nothing of the destruction that they cause. However, one can in general speak of aggression or sadism only when the individual strives to negate a correctly comprehended value. We may compare the actions of an artist who out of hate, revenge, or envy destroys the work of another, to those of a sadist who takes possession of a woman by inflicting pains or humiliations on her. Not every ''taking-into-possession'' and not every act of seizure is aggression; even less is it sadism. It depends on whether the act of taking posses-sion follows in accordance with the value and essence of the object, and in fact with respect to the subjectively not objectively comprehended essence and value.

A sadistic behavior is generally possible only when the instinctual is fitted into a totality of experience relative to certain values. The perversions are thus not derivatives of partial drives, which could also be found in the animal kingdom, but are a prerogative of human beings.

Perversions can appear only where the erotic choice is originally
determined by an orientation toward values, for example, beauty. The
perversions are directed against positive values. The cat that plays with
the mouse is no sadist because it perceives no intrinsic value—
demanding to be preserved and spared—in its booty. Thus sadism is
always a reflective and a reactive behavior, a negation of values
through the deed, that is, through the destruction of the existence of
whatever carries value. Even the so frequently observed cruelty of
children is not sadistic as long as the intrinsic value of the other's
existence and the meaning of death has not yet dawned on them.
Schilder claims, as an essential and fundamental feature of sadism, that
the sadist wants to take possession of the other unconditionally: "He
wants to be master over him and, on this account he only inflicts pain on
him in order to be assured of unconditional mastery over him."
Schilder himself poses the next question: "But does he not thus deny
the individual existence of the other? Does he not destroy him? Here
there is probably also a tendency toward destruction."[76] It should be
added, that *much more* than a tendency toward destruction accom-
panies the desire to master. The partner is by no means the same person
after the humiliation that he was before. The subjugation alters him
entirely. The exciting content of the sadistic experience is precisely
this devaluation that takes place with the inflicting of pain. The sadist
does not inflict pain on his partner in order to take possession of him;
rather degradation is the perverse form of sadistic possession. Even
Freud's claim that "at the higher stage of the pregenital sadistic-anal
organization, the striving after the object appears in the form of an
impulsion to mastery, in which injury or annihilation of the object is a
matter of indifference,"[77] misses the mark because the child certainly
still has no understanding of the values with which the objects are
endowed and can thus also not be indifferent toward them.

The desire to deduce the total mental behavior of the human being
from his biological foundations has led Freud, imprisoned in the
conceptual system of sensualism, to the disastrous construction of the
partial drives. Because of this construction the child remains a small or
partial adult in the theory of psychoanalysis; this concept has long ago
been overcome by child psychology. In order to establish this concept

76. P. Schilder, *Medical Psychology* (N.Y.: International Universities Press, 1953).
77. S. Freud, "Instincts and their Vicissitudes," *Collected Papers*. vol. 4, p. 82.

it was necessary to efface the essential distinction between the perversions of the adult and the child's modes of behavior.[78] Only in this way was it possible to succeed in identifying the "hostility toward the norm" of the perversions with instinctually decreed objects and aims. The hostility toward the norm evident in the perversions, their antagonism against the norm, does not exist merely in the value judgment of a prejudiced onlooker; rather it is the mainspring of the perversion, of the perverse experience itself. This must be vehemently emphasized. The sensual pleasure of the perversion arises only from this hostility toward the norm, from the destruction, violation, desecration, and in short the deformation of himself and of his partner.[79]

Here a short digression is in order. It seems to me to have been little noticed until now that analytic psychology interprets a certain broadly defined area of the affective life—that is, the erotic—entirely in connection with, if not in dependence on, the James-Lange theory of feeling. James pedantically formulates his principle in the following way: man does not cry because he is sad; he is sad because he cries. Thus it could be said here: the human being experiences sensual pleasure because he has reached an orgasm.[80] Actually, however, the total bodily processes of sexuality up to the orgasm stand in the same relation to the experience as the tears do to the sorrow. The direction of the sexual bodily functions depends on the experience and thus on the fact that the experiencing individual is touched by certain experiences in the very center of his soul. Experience shows that, in the more narrow sense, these situations need not even be sexual.

In anxious emissions of semen, for example, the anxiety-exciting situation brings about that central mental connection that sets the mechanism of orgasm in motion; there is therefore no necessity and no

78. The psychoanalysts, in the conclusions arrived at in their analyses, tend to see the child with the eyes of the adult, to have the adult narrate recollections from the first years of life, and to accept such recollections. The extent to which this is true is shown by a work recently published by Sadger. (*Allgemeine ärztliche Zeitschrift Psychotherapie* [1929]).

79. Cf. E. Von Gebsattel, "Über Fetischismus" ("On Fetishism"); in *Prolegama einer medizinischen Anthropologie* (Heidelberg: Springer, 1954).

80. Characteristic of this is the concept that Freud developed concerning the relation of pain to sexual excitation: "We have every reason to believe that sensations of pain, like other unpleasant sensations, extend into sexual excitation and produce a condition that is pleasurable, for the sake of which the subject will even willingly experience the unpleasantness of pain" ("Instincts and their vicissitudes," *Collected Papers*. vol. 4, p. 71). Thus, according to Freud, it is not that the erotic ecstasy arises from the experience of painful degradation and finally discharges itself in the orgasm, but that the pain as a physiological process is supposed to *cause* the sexual excitation, which is likewise conceived as a physiological process.

justification for transfiguring, through allegorical reinterpretations, a situation of this kind into a specifically sexual one. Instead the vital significance is common to both groups of experiences. The disturbances of potency, nocturnal emissions, and anxious emissions compel us to recognize that the discharge of sexual excitation is not even strictly bound to the actions and stimulations involved in normal sexual union. We may therefore legitimately presume that the sense of these stimulations and actions is not limited to their physiological being. On the contrary, they are the vehicles for meanings that conduct the experiencing of the normal individual to the same place arrived at without such mediation in the pathological case, for example, by means of anxiety. Were that not so, the already-mentioned disturbances of potency as well as that of the aspermous patient could not occur. Even the experiences of the normal individual show that the stimulation of some arbitrary erogenous zones—that is, the physiological mechanism—is not the decisive factor; rather everything depends on whether or not these stimulations are experienced as a loving caress, as tenderness, and as the nearness of a human being whom one desires to be near.

Wherever we encounter it in human existence, the erotic experience shows itself to be full of sense and saturated with meaning. Only in the region of sense are perversions possible as the negation of value-adequate erotic behavior. The perverse individual experiences violent negation and disfigurement as that central connection that sets into motion the bodily processes of sexual excitation and of orgasm. Thus the perversions can in general be understood only as reactive formations, as inversions of an eroticism bound to the sense of the experience. The theory of partial drives, and the interpretation of perversions as the original aim of partial drives, is one of the great "misguided accomplishments" of psychoanalysis. Psychoanalysis cannot comprehend the perversions because it reinterprets even normal eroticism into a merely instinctual phenomenon. The extent of the failure is perhaps nowhere more clearly expressed than in a few sentences that Freud, in his latest work, has written concerning the essence of beauty:

> Psychoanalysis, unfortunately, has scarcely anything to say about beauty either. All that seems certain is its derivation from the field of sexual feeling. The love of beauty seems a perfect example of an impulse inhibited in its aim. "Beauty" and "attraction" are originally attributes of the sexual object. It is worth remarking that the

genitals themselves, the sight of which is always exciting, are nevertheless hardly ever judged to be beautiful; the quality of beauty seems, instead, to attach to certain secondary sexual characteristics.[81]

Is it not astonishing that an investigator of the rank of Freud, after he has reflected for more than twenty-five years, passionately and with ever-renewed questioning on the problems of sexuality, ultimately has nothing further to say concerning the relation to beauty of the erotic choice—other than that it is a perfect example of an impulse inhibited in its aim and that beauty attaches to certain secondary sexual characteristics.

It would be too cheap to give a psychological—in this case a psychoanalytic—explanation for this failure. Moreover, the problem-historical and systematic contexts from which this deficiency has developed are clearly evident. Psychoanalysis transfers the vanquished and inadequate principles of associationist psychology to the region of drives. In doing so it repeats, by linking experiencing to stimulus and sensation, all of the misconceptions of the old theory. This theory has no place for a meaningful interpretation of the experience of beauty. But in human experiencing pure instinctuality is encountered as seldom as is pure sensation. Drive impulses, like sensations, are only the material of experiencing. They acquire their form, however, not through suppression under the influence of the reality principle but from entirely different sources.

Since reality for psychoanalysis is ultimately only a hindering and a postponement of direct drive satisfaction, psychoanalysis cannot comprehend the essence of the deformation in neuroses, addictions, and perversions; rather it must attempt to explain them in such a way that they seem to lie in the direction of a drive. Thus the actions and configurations (*Gestaltungen*), which we regard as aspects of self-actualization, appear to psychoanalysis to be distortions of instinctual life. For this conception Schultz-Hencke gives a pregnant definition of the intrinsic drive impulses, which have been repressed into the Unconscious:

What does the Unconscious, the "UCS", want?—To eat up everything else, to bite at others, to devour others, to give nothing to

81. S. Freud, *Civilization and Its Discontent: Standard Edition of the Complete Psychological Works of Sigmund Freud.* London, vol. 21 (Hogarth), p. 83.

others, not to work, to eat excrement, to play with it, to urinate and in general, to act when and where and how it wants; to dismantle whatever is not pleasing in its construction; to tear out and take away for oneself whatever part of a whole gives one joy; to see everything, entirely indifferent as to whether this suits the other or not; to satisfy oneself genitally whenever and wherever one pleases and with whomever one feels an impulse for; and to run away satisfied from such a situation whenever the mouth or the genitals are satisfied. And then it often has horrible feelings of anxiety. That is everything! And from the heights to the depths, what has not already been secreted away in the UCS![82]

In contrast to this, we make the claim that the cause of the neurosis cannot be sought in unconscious drive impulses; rather, the turn toward the instinctual is as such already a symptom of the neurosis and contains a deforming tendency, against which the neurotic still defends himself. We will now furnish the proof for this.

The strength of the psychoanalytic position is its foundation in the biological. But the foundation is valid strictly for the pleasure principle and is already no longer so for the reality principle, whose establishment has always been a weak point for psychoanalytic theory. Most objections have also been raised against it; up to now, however, in opposition to this biological foundation, no one has successfully suggested an alternative of equivalent value in the original equipment of the human being.

As I have already remarked, I believe it is possible to point to this alternative foundation in the *experience of time*. The experience of time is so decisively important because only when the single moment is experienced as moment, as a phase of becoming, can the contents, given one after another in the succession of moments, merge together as perspective views of the thing-world. Thus the condition of human experience is that the experiencing individual experiences himself as becoming, that is, in any single moment he also has himself only in one perspective, in one phase of his developmental passage, which is first explicated in the becoming. Thus on the objective side the experience of time founds prescientific and scientific experience, and on the subjective side it subordinates the single moment to a superordinate whole, which directs its demands to the organization (Gestaltung) of

82. Schultz-Hencke, *Einführung in die Psychoanalyse* (*Introduction to Psychoanalysis*) (Jena, 1927).

the moment. Thus the relation of the ego to experience and to duty (to the "should") develops *equiprimordially* from the experience of time.

Freud, on the other hand, attempts to deduce even the Should, the realm of freedom, from natural causality. Characteristic among many such passages is the following: "That which is concealed behind the anxiety of the Ego before the Superego, behind moral anxiety, can be stated as follows: the higher being, which has become the Ego-ideal, at one time threatened castration, and this castration anxiety is probably the nucleus around which the later moral anxiety settles; it is this that continues as moral anxiety." An external impediment to direct drive satisfaction, through identification and the reception of the command, produces the conscience. Reality is thereby always comprehended only as resistance; the command is originally a prohibition.

The concept of identification does not suffice, however, to explain the phenomenon of the conscience, the transition from the can-not to a may-not. Why does an individual comprehend the resistance of a wall, when he wants to pass through it with his head as a can-not and by no means as a may-not? Why does he not identify himself with the wall? Identification with the father, however, only leads to the formation of conscience if his commands are viewed as an exercise of his right and not as an expression of his might and if the father is regarded as person, as representative and administrator of right and justice. Even the small child already distinguishes very precisely between those commands that emanate from caprice, mood, and malice and those that are issued for justice' sake. Children are very sensitive toward injustices, even when there is enough real might behind the unjust orders to carry them out.

The Should is originally a behavior turned against itself, the expression of the subordination of the single moment under a whole, which explicates itself in this moment. The Should establishes in general and for the first time the "may" and the "may-not." Because one should, one also can ask what one may or must do in the individual case. The educator, the adviser, the system of ethics, statute law, religion—all give answers to this question. They can only give answers, however, because they are asked, just as empiricism can answer only because what is given in the moment is already, for prescientific man as well as for the primitive and the child, experienced with reference to the incomplete system of experience and knowledge. The historical metamorphoses of statute law and the multiplicity of ethical systems do not annul the idea of justice, any more than do the conflicts among discoveries and the advances of scientific knowledge annul the possibility of knowledge in general. The prohibition does not hover before an action and make a guilty deed out of one that is neutral in itself; the action is experienced from the outset under the aspect of the Should, under the demand of an individual idea. The bad action shows itself to be bad in itself, and by this means becomes, individually or in an historical development, a forbidden action. The Should exists originally as the positive precept, as command and prohibition. Just as human knowledge elevates itself above instinct and training, so too human action extends beyond drive and external compulsion.

The moral demand is thus by no means merely carried over to the individual, through the fact of social relationships, from the outside as it were. Because ethical behavior is a behavior originally corresponding to the experience of time, it stands just as close to, or just as far from, the biological foundations of the human soul as do sensation and perception. The psychology and psychopathology of perception may have gotten a head start over the psychology of feeling and the psychology of the value-experience for reasons of research technique, but it has no methodological claim to any priority. Thus there is also no basis for the criticism that, by discussing such questions, we stray further and further from the problem-domain of medical psychology. On the contrary, only by this means will the problems of psychopathy and of addiction be fully disclosed.

The subordination of the moment under a demanding whole has as its consequence a *devaluation* of the moment. The moment is robbed of its absolute, intrinsic value; its value is now determined through its relation to the whole There now arise two movements opposed to one another. The one aims at the fulfillment of the demand proceeding from the whole—through *objectification*, through production in the work, through having an effect in the medium of social institutions, and through the creation of the individual figure (Gestalt). We call this direction self-actualization. It reclaims for the moment its intrinsic value, not by reference to another moment, which certainly in itself can have no higher rank,[83] but rather by reference to the individual figure (Gestalt), removed from the temporal relation. Folk psychology teaches that in the most primitive groups, the horde has an entirely

---

83. That would be the function of the self-preservative instinct.

different structure than the herd; it is by no means a mass of individuals held together by instinct, it creates its social institutions as a medium for self-actualization. Similarly, human motherhood is not simply a continuance or descendant of an instinct to nurse one's brood. As to the original natural man, so highly praised by many biologizing psychologists, the natural man whom the culture has not yet corrupted or alienated from himself, he is found among ourselves, actualized in the many forms of high-grade feeble-mindedness.

The impulses adjudged to primal man by psychoanalysis are neither primordial nor instinctual; they are in reality markedly reactive, and their tendency toward deformation directed against self-actualization is already characterized by this fact. The coprophiliac has no original joy at touching the feces; rather the meaning of the coprophilic experience develops in him out of the significance of fouling, profanation, and self-destruction. Flies that look for nourishment in feces are not coprophiliacs, no more than are chickens who scratch in manure, or patients who take diuretin prepared from guano. Still less is the child who plays with his own excrement a coprophiliac. An astonishingly crude mode of observation is required here to overlook the decisive difference between the behavior of a child and that of an adult, and to fail to recognize that the child and the adult construct objects of an entirely different nature from the same material. For the coprophiliac the feces are not an object characterized by its physical properties; rather, he aspires to have contact with the excremental, the putrefied, the decayed. A small child can play harmlessly with his excrement, harmlessly because he has not yet understood decay and therefore does not yet feel disgust. In contrast, the experience of the coprophilic derives directly from the union with what is disgusting.[84] Such a reactive moment is inherent in coprophilia as well as in all perversions. Thus, as we have already shown for sadism, it is precisely because of this reactive, deformative character that they can never have been original instinctual aims.

The movement opposed to self-actualization attempts to flee the demand proceeding from the whole. Nevertheless it is just for this reason that this movement remains bound to the whole and related to it.

---

84. Shame and disgust are not the result of educative measures applied to the child from the outside; rather they are vital modes of behavior directed against that which is decaying. I expect to demonstrate this in more detail elsewhere. (Cf. E. Straus, *On Obsession: A Clinical and Methodological Study* [N.Y.: Nervous and Mental Disease Monography, 1948].)

In each of its individual phases it is dependent on and derived from self-actualization. It can attain its aim of self-abandonment, of "letting oneself fall into decline" in all of its degrees up to self-destruction, only through the deformation and destruction of the forms and structures serving self-actualization. Self-abandonment cannot take aim directly against the demands of self-actualization; it can accomplish this only by turning against the works, institutions, and social forms that serve it. It is precisely because of this indirectness, however, that it cannot get free from the demand it seeks to escape.

In the effort to reclaim the value of the moment, we encounter another possibility for escaping the demand of the whole. Festival and celebration, frenzy and intoxication, pleasure and debauchery collectively serve this purpose. But even here we discover mainly an intermingling of deformative tendencies. As Goethe's description of the Roman Carnival in his *Italian Journey* shows, and as the Rhine Carnival or the tumult on Sylvester's Eve can still teach us today, the ridicule of customs and institutions, of individuals in high offices, and of personalities distinguished above the crowd as well as the caricature of the beauty and dignity of the human figure form an important constituent of all such festivities. At the same time the freedom inherent in wearing a mask makes it possible for the participants in the festivity to allow themselves to carry on at will, now in this direction, now in that direction, to approach strangers under the protection of anonymity, and to give themselves over to the hurly-burly. Thus great crowds of people are a part of all such events, as is confusion similar to that of the waves of people in a dance hall, where the individual dragged into the general movement now leads, as it were, only the existence of a molecule and devotes himself entirely to the moment in order to consecrate the moment. Nevertheless the demand still does not become entirely silent; it is merely temporarily suspended and remains in the background of the festive events. Thus the day and the hour, the beginning and the end, are prescribed for this event in advance. In the midnight hour of Shrove-Tuesday Eve the carnival goings-on and the right of the mask suddenly cease all at once. This sharp interruption, this sudden return to the customary order, expresses the fact that even the time of license and freedom is contained by the encroaching, collective form of the objectifying tendencies. The need for purification, light, clarity, and order announces itself. Everything has been only an apparition, and everydayness lies before us once more in a new fullness. Festival and frenzy are not only important as the illusionary success of the suspension of the demand; they also lead back to its affirmation and acknowledgment.

Only the addicts struggle beyond a temporary suspension toward a radical dissolution of the demand. They desire to linger and become absorbed in the pure objectivity of the moment. Through the multiform means of intoxication, they procure for themselves their "artificial paradise." As manifold as the addictions are, they nevertheless have this decisive factor in common—the attempt at lingering in pure objectivity. Addicts do not seek forgetfulness in pleasure; rather the pleasure resides in the forgetting. The morphine addict, for example, produces in the euphoria a pleasure-accentuating bodily state. In morphine intoxication the body is no longer experienced as it was previously as the organ of orientation toward the world. The euphoria makes it possible to linger or to sink into absorption in the pure pleasure of bodily existence.

Therefore, even without any toxic compulsion, addiction leads to repetition. It must do so because, with the fading away of the euphoria, the summons again becomes audible—the moralizing and the "Katzenjammer."[85] Even the hangover is by no means a purely bodily conditioned experience; in it the corporeal misperceptions are joined by the feeling of breakdown in the face of the reawakening demand and by the impression of having failed generally to silence the demand. Thus the burden and the torment resulting from the demand are even more difficult to bear in the fading of euphoria than previously and impel one now more than ever toward a repetition of the euphoric state. Thus degeneration and depravity in the attitude of the addicted are present from the outset.

These few remarks cannot exhaustively present the multiform symptoms of the addictions. They are intended to provide only a schema, which should suffice to make the deformative tendencies in the behavior of the compensation-neurotic intelligible. He too seeks to escape the demand for self-actualization and to sink into a purely vegetative existence. One is almost tempted to speak here of the addiction of those without the means of intoxication. This addiction is manifest in the few who, after the award of a pension, continue to live

---

85. *Katzenjammer*: idiom for the qualms of conscience and resolutions to change one's ways that occur during a hangover. (Translator's note.)

in a psychogenic isolation and confinement. The others are driven by it into the compensation-proceedings but find themselves again as soon as the proceedings are decided in the positive or negative sense. It is conceivable that such deformative tendencies toward self-abandonment lie in readiness in many who lead their life in the form of proletarian existence because here self-actualization is arrested from the outside; even their active life is—viewed historiologically—[86] in many cases only a matter of running in place. In its ultimate import it is fundamentally no different from compensation-neurotic self-abandonment. External life conditions, like all other genetic factors, do not cause but rather merely activate the deformative tendencies, which always continue to presuppose the summons of self-actualization. Secure abundance can thus give rise to self-abandonment just as well as does need. The compensation-neurosis is thus certainly not restricted to those who must get by with the minimum in existence; the well-to-do can likewise fall under its power. In the group of well-to-do compensation-neurotics, perhaps only the bad-intentioned malingerers who seek to exploit ruthlessly an economic chance are relatively more numerous. Next in frequency are those inclined toward complete self-abandonment, and they closely resemble those among the needy who hold fast to a spurious form of life. The various manifestations of addictions vary with the external conditions of life, just as all addictions carry their specific imprint. At first glance the differences between someone vegetating on a scanty pension and a blasé globetrotter pursued from amusement to amusement by the spectre of boredom, will be more apparent to the eye than the commonalities. Nevertheless it seems to me that these figures, so distant from one another, are, like all addictions in general, variations of the same basic behavior of the *deformation*. The compensation-pension thus does not serve the security of existence as long as it still contains traces, residues, and possibilities of self-actualization; instead it makes possible for the first time the full self-abandonment into pure vegetating. The battle for the pension may certainly be fought tenaciously and with great energy; yet measured against self-actualization it is mere idling.

---

86. *Historiologically*: Straus defines the historiological perspective in the context of his discussion of shame. This term denotes that human actions, experiences, and expressions can be understood only "with the help of historical-psychological categories. . . . This is so because the experiencing person understands himself historiologically, i.e., as continuously emergent." (Straus, E.: *Phenomenological Psychology: Selected Papers* [N.Y.: Basic Books, 1966], p. 217). (Translator's note.)

While the morphine addict, for example, can artificially produce a euphoria, that is, can be consumed in a state of heightened pleasure, the compensation-neurotic does not have such means at his disposal. He might wish to turn toward a purely bodily existence, but when he does this or attempts this the general mode of givenness of the body is altered in the experience. As soon as the body itself comes into givenness in the experience and is no longer the organ of our orientation toward the world there is an alteration of the common, normal feeling. From the indifference develops a discontent. The compensation-neurosis removes the bodily event from indifference, and because the pensioned individual does not have the means at his disposal to activate euphoric processes, the bodily processes must come to consciousness for him with an accent on unpleasure. The fact that he conducts himself histrionically, however, does not yet determine whether he experiences the body, or the bodily events, in a manner deviating from the normal. There are certainly also organically injured individuals who do not believe they can get by without the help of histrionics. Histrionics would be mere malingering if there were no change in the mode of "having" the corporeal functions.[87] In the beginning of the development of a compensation-neurosis the histrionics may be even more important than the altered "having." Nevertheless, the more the deformative tendencies of self-abandonment gain in efficacy, the more the modes of experiencing the bodily functions, if not simultaneously the functions themselves, must also be altered.[88] These principles concerning the manifestations of the compensation-neurosis, which the experiences of everyday life already teach us and which we have formulated here theoretically, have also been verified by experimental psychological investigations undertaken by Lewin et.al.[89]

In investigating psychic satiation, Karsten had subjects carry out activities of various kinds, such as sewing, transcribing, sketching, with the instruction: "You are to work just so long, until you've had enough." If satiation was reached, then along with a gradual deteriora-

87. In the experience of illness, the individual "has" his body in a different modality, for example, the individual "has" symptoms. With this altered mode of "having" the body, physiological functioning may or may not undergo a corresponding change. (Translator's note.)

88. Cf. also in this context my work, *Wesen und Vorgang der Suggestion* (*Nature and Process of Suggestion*) (Berlin: 1925). (To be forthcoming in a later volume from Duquesne University Press.)

89. *Die Entwicklung der experimentellen Willenspsychologie und die Psychotherapie* (*The Development of the Experimental Psychology of Volition, and Psychotherapy*) (Leipzig, 1929). See also Karsten, *Psychol. Forsch.* 10 (1928).

tion of the performance, there also emerged tendencies toward
abridgement and shirking as well as a sensation of fatigue. It can
nevertheless be demonstrated experimentally that corporeal, muscular
fatigue was not the real cause of the process of satiation. The "not-
able-to" feeling, like the muscle pains, was somehow not simulated,
but was thoroughly genuine. The experimental subjects experienced
difficulties in carrying out the activities as a failure of their corporeal
performance capacity and sought to compensate for them through
heightened exertion. In spite of the pronounced strain of the subjects
their failure did not lie in muscular exhaustion. They were again able to
execute immediately and faultlessly a new task even when the new task
was composed of the same action-components as the old. Lewin
conceives the tiredness in these experiments as a corporeal symptom
of psychic satiation and equates these experimentally induced
phenomena with hysterical phenomena. He emphasizes as a crass
example the behavior of an experimental subject "who in the first hour
of the experiment became satiated first with sketching and then with
reading poetry." In the second experimental hour she became markedly hoarse while sketching. The experiments show the extent to
which well-being depends on the task to which someone devotes
himself. I concur with Lewin that no simulation was present in these
cases, but on the other hand it seems to me that this conception, that the
tiredness is a corporeal symptom of psychic satiation, misses the point.
In psychic satiation there is no coming into play of new corporeal
functions (and organ-sensations dependent on these functions) unlike
those present in the beginning of the experiment; the same functions
and sensations are experienced differently in the beginning and at the
end of the experiment. With the alienation of the individual's interest
from the work, the "doing" comes into a different light as a "must-
do." We have already discussed above the fact that precisely in those
instances where one wants to experience one's own bodily motion, it is
still necessary to transform the activity into an action, to give it form
and aim such as through the rules of a game or through the figures of a
dance or of a gymnastic movement. Pure movement does not alone
suffice, not even if the body were experienced in its own motion. In the
intellectual realm too an exciting discussion can immediately again set
in motion a train of thought that previously had come to a standstill
with a sensation of tiredness, of "not-being-able-to-go-further." The
emergence of a goal eradicates the sensation of heaviness and diffi-

culty, while, inversely, with the abolition of a goal bodily existence
presents itself to the experiencing individual in the sensation of heaviness and burden. The organ sensations, like the sensory sensations, are
material only for our experience of well-being or ill-being. In order for
our well-being to change into ill-being there is absolutely no need for
different corporeal processes and sensations to be present; an alteration
or an abolition of the goal and the ending or the slipping-away of
the possibility of self-actualization already suffice. In the self-
abandonment that sets in from the beginning of the compensation-
neurosis, the bodily state of mind (*das leibliche Befinden*) changes
from indifference or well-being into ill-being, but this need not compel
us to assume that different corporeal processes occur or that the
complaints of the compensation-neurotic are mere histrionic display or
even malingering. The histrionics can in fact resolve nothing concerning the altered ''having'' of the bodily states. In accordance with what
has been said here, I am convinced that the compensation-neurotic
individual finds himself ill-disposed as a consequence of the self-
abandonment and that his histrionics only give expression to this in a
heightened fashion. Thus there is something spurious in the exaggera-
tion of the expression, but not in the fact that troubles would be
simulated without entering the individual's awareness. The compensation-neurotic perceives troubles as soon as he abandons himself,
only not in the same measure or with the same scope as he histrionically
displays them.

In this situation self-abandonment remains continuously linked to
the pension; because in self-abandonment one certainly does not wish
to give up life in general, the individual suffices himself with narrowing
his life to a vegetative existence. The self-abandonment rests precisely
in the fact that the compensation-neurotic renounces work and wage
and confines himself to the pension. So far as one speaks of a tendency
toward ensuring security in the compensation-neurosis, that which is at
stake is only the security of vegetative existence. What is remarkable,
is the extent to which the Social Security Insurance and the Social
Services themselves harmonize with this compensation-neurotic attitude. They set up a tremendous apparatus in order to safeguard a
purely vegetative existence, and in this respect they harmonize absolutely in their estimation of the meaning and value of this vegetative
form of existence with the estimation of the compensation-neurotic.
One sees how the institution of Social Security Insurance and its

beneficiary, the compensation-neurotic, are both creations of an historical development, so that it would not be illegitimate to call the compensation-neurotic individual an unguilty culprit. It thus also remains questionable, whether one should necessarily infer a psychopathic constitution[90] from the mere presence of a compensation-neurosis.

The concept of psychopathy generally presents a characteristic difficulty, namely, on the one hand "psychopath" is understood to refer to a biologically variant constitution, but on the other hand the deviation from the biological norm is not measured exclusively in biological functions; the standard in its essentials is formed by cultural and social demands. However, adjustment to certain forms of cultural and social structures, all subject to historical mutation, is still no standard for biological givens.

We may not devalue as psychopathic every human being who does not fit the cross-sectional demands of our era. The refusal of the forms of life handed down to him, the secret or open revolt against them, and the tendency toward self-abandonment do not permit one without further ado to infer a psychopathic constitution.

However one may be inclined to assess the constitution characteristic of the compensation-neurotic, the renunciation of self-actualization that serves as the basis for his behavior is already presupposed by the traumatic experience. This renunciation does not arise in the accident-incident and does not result from it through a compulsion toward sense-derivation. The insurance obligation of the accident merely makes possible the continuance of that renunciation. There is a compulsion toward sense-derivation only in those shocks that take place in the accident-experience, which later find their expression in phobic and hypochondriacal reactions.

Thus the praxis of expert assessment will not be altered by our efforts to gain a deepened insight into the structure of the psychic trauma and, by means of the psychic trauma, into the psychology of the compensation-neurotic.

The special place of the phobias has already been acknowledged in recent general discussions of the compensation-neurosis. Theoreti-

90. In German psychiatric terminology *psychopathy* and *psychopathic* have a more general meaning than in our own. They designate any psychic pathology, any disturbance of affect or behavior that can be attributed to an aberrant inborn disposition (constitution). (Translator's note.)

cally our analysis may contribute toward eliminating difficulties and doubts. Diagnostically, new practical problems emerge once we acknowledge the special place of the phobias, because as soon as it has been bantered about in interested circles that there is a possibility of attaining the sought for pension by means of these phobias, many will hurry toward this new asylum. That, however, should not hinder us from holding firm to the distinctions concerned.

The diagnostic difficulties become especially great when originally genuine phobic reactions are built into deformative, compensation-neurotic behavior. These difficulties in diagnostics, however, must not deter one from holding firm to the distinctions laid down here for expert assessment and for treatment.

If one is dealing with phobias arising from the shocking experience without any overlay, then the effort to influence them psychotherapeutically will again reveal the vast difference separating phobic reactions from the "deformative reactions" involving self-abandonment. The psychotherapy of the simple phobias has the same aim as the cathartic procedure customary in the early days of the development of psychotherapy, and still applied today by Frank and others. Even though therapy has developed beyond this initial stage, it has nevertheless not abandoned its original position in emphasizing the individual experience and the continuing efficacy of its pathogenic power. In catharsis the patient, through hypnosis, is supposed to again live through events that are either freshly present to him in the beginning of the treatment or only gradually resurface in hypnosis. The process of living-through-again and the subsequent process of talking-things-through are supposed to eliminate the existing disturbance. Only for a small number of cases, relatively simple in their makeup, does the result live up to the expectation. I am not satisfied with the theory interpreting this process as an abreaction of affects. When the cathartic procedure does lead to a result, this result has been reached not because some kind of psychic energies have been discharged but because a past event has gained a different meaning for the experiencing individual. This transformation consists in the fact that a representative event becomes a banal event. To return to our original example, the meaning: "the human being can die," as the youth had first experienced it, makes way for the meaning: "a human being has died." The traumatic experience then appears merely as the actualization of one possibility among the many, which could have occurred.

PART TWO

The Archimedean Point (1957)

8 / The Daytime View and The Nighttime View

In the statements of physics the image of the world so familiar to us in everyday life—the daytime view—is replaced by the nighttime view,[1] that is, by a system of nonobservable, mathematical formulations. Physics is therefore intent on eliminating all anthropomorphic qualities; yet along with the anthropomorphic, physics is ultimately threatened with the disappearance of man himself (Anthropos). Physical reduction can certainly not at some point call a halt and spare the physicist alone. Thus in its practice this rigorous science does not adhere to its own principles. Actually the physicist, who develops this nighttime view for us, never leaves the secure ground of the everyday world; no more than the fakir who before the eyes of his astonished audience appears to climb aloft on a rope swinging freely in the air.

Fechner believed it is possible to reconcile the daytime view with the nighttime view by a kind of Spinozistic pantheism. It is no doubt legitimate, however, to attempt an anthropological solution before a theological solution. The observer—who sees colored things, dispenses with them, and interprets colors as waves—makes the separation and the unification of the two views possible. The separation precedes the unity; the unification is basically a reunification. In natural science, to be sure, once the separation is accomplished the possibility of unification is often forgotten or even disputed.

Nevertheless, the physicist, to whom we are indebted for the formulae of the nighttime view, remains a citizen of our human world; all his scientific activities are carried out in the everyday order of visible, audible, tangible objects. His descriptions and communications are

1. G. T. Fechner, *Die Tagesansicht gegenüber der Nachtansicht* (*The Daytime View versus the Nighttime View*) (Leipzig 1879).

143

bound to human language and directed to his fellow human beings. The certainty of laws in physics points back to the reliability of everyday experience. Like all human accomplishments, the kind of knowing that occurs in physics also requires an acknowledgment of the everyday world in its macroscopic form. Nevertheless, there are many who expect a definitive explanation of all human and animal behavior to arise from the investigation of the brain and its microscopic structure and function.

Scientific method decrees as self-evident that one begins with the known, or knowable, and pushes on from there toward the unfamiliar. In neurophysiology this rule is not always followed. Much zeal has thus been involved in inventing models of the brain that would make possible a replica and explanation of behavior and experience. Without concerning oneself much with the structure of the original one begins with a reconstruction. This forcible attempt to express behavior and experience in the conceptual language of the cerebral mechanism is encouraged by two illusions, both of which are natural to us. We are inclined to interpret as knowledge our familiarity with the everyday world as well as to find in the effortlessness of our mental accomplishments a sign of their simplicity. Indeed we overlook the complex structure of our apparently simple experiences.

In his experiments the physicist takes for granted the possibilities of seeing, of observing, and of describing—as we all do in our daily dealings with humans and things. For the psychologist, on the other hand, the seeing of what is seen, the observing and describing of what is observed, and the thinking and interpreting of what is thought and interpreted become problems themselves. The theme of the psychologist is the seer, the seen, and the seeing. In the role of anthropologist, man is simultaneously the measurer and the measured. However, in a certain way the measurer must be mightier than the measured, the limits of which he determines. Only he can determine the limits because, stretching beyond the limits, he dispenses with the factual by means of the possible. It is therefore highly doubtful that the measurer and the measuring can be explained by reference to the measured.

How is it even possible to relate "human behavior" and "the brain"? In this case the little word "and"—easily said and quickly written—unites two essentially different themes. Behavior is the true element of our existence from the first day of our life to the last. Human behavior belongs to our everyday world of colors and sounds, of macroscopic dimensions, of "natural" sizes. In acting and reacting we all become familiar with that; no apparatuses and no experiments are necessary. The brain and its anatomy and function, on the other hand, are the domain of work of a small number of scholars. Its investigation occurs under unusual, indeed abnormal, conditions: on the operating table and the dissecting table. More precise study requires the application of the most complicated instruments. The field of cerebral mechanisms is a universe of bare quantities, impersonal processes, microscopic structures, and molecular and atomic dimensions. It is a world in which light does not illuminate, does not brighten the room, nor make any objects visible to the seeing eyes. In order to compare two such different regions and to relate them a mediation is required. It is the observer who functions as mediator, who in the reality of the everyday world observes the behavior of his fellow human beings. Precisely there he occasionally comes across a brain and investigates it in order finally to convert his perceptual observations into abstract mathematical and physical concepts.

The physiologist, who in the everyday world relates behavior and brain, actually makes three kinds of things into objects of his reflection: behavior, the brain as macroscopic formation, and the brain in its microscopic structure and biophysical processes. From the whole—the living organism—the inquiry descends to the parts: first of all to an organ—the brain—and finally to its histological elements. Statements concerning the elementary processes acquire their proper sense only in reference back to the original whole. The researcher is able not only to observe the whole but also to interpret the elementary event. This personal union of observing and interpreting makes it possible to connect thematically three such different sorts of objects as behavior, brain, and cells. Physicalism in psychology, however, believes it possible, ultimately, to eliminate the observer; it presumes that the accomplishments of the observer may be directly attributed to the last member in the sequence—the microscopic structures.

In the technical language of physiology the sense organs are characterized as receptors. The organism is understood as stimulus receiver. The relation of the seer to the visible world is interpreted as a relationship between light rays and a light-sensitive sensory surface. Physiology views the eye—correctly—as an optical instrument. Physiology's methods become questionable only when they seek to conceive the seeing and the seen, visual space, and the objects of seeing entirely as accomplishments of this apparatus and to construct them as an out-wardly projected—or actually rejected[2]—image. Problems crop up within this conception, problems that cannot be surmounted by purely physiological concepts, above all problems concerning the phenomena of the unity of visual space, of constancies, of "depth," and even of objectivity and of seeing itself.

In the study of the brain's performance, neurophysiology makes use of certain apparatuses. The organ, in its unfamiliar, yet-to-be-investigated functions, has an effect on machines, whose physical performance is familiar. From the variations of electrical potential appearing in a measuring apparatus, the processes in the brain that is producing such variations are inferred. Thus the concept of sensory performance is narrowed in a remarkable way. Indeed one knows that the calcarine cortex has something to do with seeing. Nevertheless its characteristic function—which is appropriate to it as an organ of an experiencing individual within the whole of the I-world-relationship—remains disregarded in the physiological analysis. Instead the specific nature of the cerebral function is defined by the physical characteristics of an apparatus. It is interpreted as an impersonal event localized at a definite place; the seeing and the seen are narrowed down to eyesight and are classified along with measurable processes in the *area striata*.

Democritus is said to have taught that little images, *Eidola*, are released from visible objects and penetrate the eye of the seer. This concept appears primitive to us. As physiologists of the senses we interpret the visible object as optical sender and the eye as optical receiving station. Then indeed we again come suspiciously close to the image theory of Democritus and interpret the seeing of the visible as the appearance of images in consciousness—a kind of positive imprint of the calcarine-negative. The perceived object is reinterpreted as a per-ception that is supposed to be grafted onto cerebral processes in some mysterious manner. "All *sensa* are identical with parts of the brain and not with the surface of the external physical objects," writes Price[3] in a book on perception.

The consideration of the mediating role of the observer certainly

---

2. *Rejiziertes*. Straus draws on the etymological meaning of *projizieren* (English, "to project"; Latin, *pro-jacere*) as "to throw forth," and creates a new form *rejiziertes*, meaning "thrown back again" or "rethrown" (translator's note).

3. H. H. Price, *Perception* (London, 1950).

makes the task of the neurophysiologist more difficult, but at the same time it clarifies the situation. As observer I relate human or animal behavior to the brain. This comparison is my work—one selection out of the totality of possible human behaviors. Accordingly, it is necessary to include scientific observing, describing, and comparing in the region of things to be explained neurophysiologically. In my environment I, the observer, meet human beings who, like myself, are capable of establishing relationships to visible things and to other human beings in their surrounding world. The human as object of my observation must himself be understood as observing subject. Just as it is now my brain that enables me to see, to observe, and to describe, so it is his brain that renders him, my fellow human being, capable of similar achievements. Behavior and experience are constantly my, your, or his behavior and experience; and they stand as such in relation to my, your, or his brain.[4]

The physiology of the brain does not dwell long on these basic relationships. It ignores the possessive-relationship; it replaces— generally without giving an account of it—my, your, or his brain with *a* or with *the* brain. The physiologist probably has no other choice. Nevertheless the reference to the possessive relationship may not be dismissed as a sentimental claim. It is important to look into and acknowledge the unavoidable circumscription of the research results accompanying that substitution, because the elimination of the possessive relationship distorts the phenomena, narrows down the problem area, and thus tacitly anticipates a theoretical judgment. If my, your, or his brain is replaced by *the* brain, then the brain is generally viewed not as an organ of an experiencing being, but rather as a steering apparatus of a movable body. The observer undertakes the physical reduction of the world that is accessible to him in his scientific experience. He places the brain in this physical system as a corporeal formation acted upon by other bodies and reacting to them. In the end the observer is surprised to notice that processes in the brain are at times accompanied by processes of consciousness. Presumably he now endeavors to define and to conceptualize this puzzling addition of a consciousness.

In every anatomical and physiological observation of the brain, two

---

4. Cf. R. Hoenigswald, *Die Grundlagen der Denkpsychologie* (Foundations of the Psychology of Cognition). 2nd ed. (Leipzig-Berlin, 1925); L. Binswanger, "Der Mensch in der Psychiatrie" ("The Human Being in Psychiatry") *Schweizer Archiv für Neurologie und Psychiatrie*, 77 (1956), among other works.

brains are involved: the brain of the observer and the observed brain.[5] The elimination of the possessive relationship compels one to ignore this fundamental fact. If it is legitimate to replace my, your, or his brain by *the* brain, then both brains must in principle be exchangeable— notwithstanding all differences of kind, constitution, and life history. The spatial and temporal relations that are valid for the observed brain must also be definitive for the observing brain—and finally for the observer himself. A confrontation of the observed brain with the observing brain enables us to make a test case of the example. If it is possible to trace behavior back to the cerebral mechanisms, then one may justifiably demand that the observed brain, too, must carry out all the accomplishments of the observer and the observing. In such a comparison it becomes apparent that in the usual execution of the reduction not only is experience from the beginning reinterpreted in favor of mechanisms but also unnoticed anthropomorphic achievements have been attributed to ''the brain.'' The violence in the way behavior has been treated finds a necessary compensation in an anthropomorphic interpretation of the brain. One grants less to behavior than is due to it, and gives the brain more than belongs to it.

5. While the observer studies a brain that is not his own, in a living organism or in an anatomical preparation, his own brain has been excited by it through stimuli. While the observer communicates findings that he has gathered from the brain of a corpse, or of an ape, dog, or rat, it is his own brain, his calcarine cortex, that has been affected.

9 / *Dimensions*

In macroscopic consideration and microscopic dissection, the brain is assigned to Euclidean space, that is, a homogeneous, isometric, and isotropic[6] three-dimensional continuum. The space of the observer, on the other hand, is not homogeneous, isometric, isotropic, or continuous.[7]

The space of the observer is not homogeneous, rather it is divided into near and far zones: the constancies dominating in the near disappear in the distance. It is not isometric: parallel lines tend toward a vanishing point in the distance. (In nearness, on the other hand, parallels can be constructed in accordance with Euclid's postulate.) It is not isotropic: while horizontal parallels appear to converge, the verticals do not follow this transformation—an indication of the role that gravity and its overcoming play in the formation of sensory space. It is not continuous: a tension exists between the Here—the momentary abode in which each individual is contained in his corporeality and gravity—and the many Theres, which show themselves to us in our mobility as many possible close and distant goals.

Psychology deals with human beings and animals as experiencing beings. Thus it deals with experiencing beings insofar as they are human beings and animals, that is, creatures who are able, who are compelled, to provide for themselves, to seek and to find, and to attack and to defend in pursuit of the "means of life"—the nourishment necessary for their life process. Sensibility and motility stand in the

6. Isotropy holds only for purely geometric, and not for physical space. Insofar as the brain is viewed as a corporeal formation in a gravitational field, one cannot speak of isotropy in the strict sense.

7. We limit ourselves to the world of visible things but do not forget that visual space is only one mode of spatiality. Certainly, and for essential reasons, this visual mode has a priority above other spatial forms for human behavior and observation.

service of these life processes. Animal and human locomotion are significant only in relation to nonhomogeneous space, in which the goods of life are unevenly distributed. Locomotion arrives intention-ally[8] at a place that is preferred over the momentary present abode. Intentional movement is determined by the topography of goods.

Visual space is neither external space nor a space of contemplation (an observed space). It is not a panorama on which the seeing individual looks as though from the outside. In seeing, just as in all sensory experience, we experience ourselves in a relation to the world. Seeing is to be understood as a mode of our natural "being-in-the-world." The total content of the relationship of seeing to the world—a relationship unfolding in the acts of seeing—can accordingly not be represented by a reproduction of visual objects alone.

In space and together with things, we find ourselves nevertheless in contraposition to (*gegenüber*) everything else. Being in contraposition (*Gegenüber-sein*) is an irreducible relationship for the experiencing individual. It is founded in mobility, not in the completed motion nor in the circular *Gestalt* of perception and movement but rather in the necessity accepted along with the possibility of overcoming gravity, of freeing oneself from the earth, and of behaving, in the tenacity of corporeal existence, in other words, in opposition to everything else. The possessive relationship is a mode of contraposition since only in opposition can that which is mine be contrasted with that which is other and distinguished from it without being separated. Only a mobile being can sensibly comprehend the other as other.

In the basic Cartesian concept of consciousness, the *res cogitans* is addressed as "I." For its experience of itself the ego requires no "thou" or "it." It does not relate itself to others or to what is other. In the loneliness of its worldless existence, the I of consciousness notices that, in the world of things with extensiveness, one specific body is quite intimately bound up with it. It recognizes this body as its own. The bases for this judgment are furnished by the sensations that intrude through the body upon the soul, and above all else by pain. Descartes said that we sense pain *in* our body. According to this, the possessive relationship would originally be the relation of an intramundane body to a soul already individuated before this connection. It operates as an

---

8. With a hyphen Straus emphasizes that the German *Ab-sicht*, or intent, derives from *Sicht*, or vision. *Ab-sicht*, then, means seeing away from me, seeing what is now at a distance from me, that toward which I can move (Translator's note).

infringement of free will and clear judgment. "Mine" would be the body that confuses my thinking and holds me in the region of sensations and passions. Nevertheless, the possessive relationship "my body" expresses an innerworldly relationship. By raising ourselves against gravity and holding ourselves erect over the supporting earth, we differentiate ourselves in our mobility from the surroundings. I do not experience my body as an object somehow outstanding among other objects; rather I experience the world in my embodiedness. The possessive relationship envelops antithetically that which is other, that which I am not—the objects of my acting and experiencing. I am nevertheless bound to this not-I, from which I differentiate myself as an embodied being. Boundary-surfaces and boundary-lines separate only that which can be in contact within an encompassing whole. Impressions of touch are simultaneously a touching and a being touched. The skin is the boundary-surface of the body. I see my hand just as I see any other object; my own voice can sound strange to me; in touching I experience myself as touched upon directly. Contraposition is therefore not originally or exhaustively an optical phenomenon. It is not limited to the visible, although perhaps it finds its purest expression in the frontal alignment of the "straight-forward" stance. Thus, for the undistorted comprehension of an object, we are also inclined to grant priority to the orthoscopic nearness-attitude above all other attitudes. In relation to such a confrontation we then speak, with dubious justification, of the true size of an object.

In our bodily existence we are limited to a minute region of space. Only an arm's length from the saving shore the exhausted swimmer is snatched away by the tide; by a hair's breadth we escape a deadly collision. Distances are unrelenting. Nevertheless, the outer bounds of our body are not boundaries; space unfolds before our eyes in its entire breadth. In the relation of contraposition that which is other becomes accessible in its entirety to the seeing individual.[9] With respect to the totality of the surrounding space we are able to comprehend our Here

9. In *Being and Time*, Section 70, Heidegger says: "Dasein is never present at hand in space, not even proximally. . . . Dasein takes space in; this is to be understood literally. It is by no means just present-at-hand in a bit of space which its body fills up." (p. 419, English edition) Since Heidegger takes his terminology from the essential elements of the German language, one faces a certain embarrassment with statements that only occasionally go beyond the region of the ontical-descriptive. Apparent similarities result that do not always indicate agreement, and deviations that do not necessarily signify contradiction. We are indebted in ways that are particularly difficult to demonstrate, yet which nevertheless must be summarily acknowledged.

—the abode in which we are momentarily held by gravity—and as mobile beings are able to determine our location in relation to other places. All spatial determinations, practical as well as theoretical, can be accomplished only in descending from the whole to my own position. Thus, comprehending my place in relation to the whole of space bounded within a horizon, I can move myself from a start to a finish. Space unfolds itself before me as a space of action. The There lies as the endpoint of a possible action—visible and graspable in the present but nevertheless distant and futural as a goal before me. From my center outward the world around me arranges itself for me into zones of near and far goals. In it things appear alluring or offensive, soothing or shocking. I experience things in relation to myself in just such a physiognomic aspect.[10]

In contraposition to all other things, I find myself constantly in the center, although certainly not in the strict, optical-geometric sense, such as when a circular horizon surrounds us on the high seas. The manifold forms of limited space preclude speaking of a center in the geometric sense. Yet space is also centered for me within our four walls. It opens itself before me and for me in a manifoldness of direction, articulated by the principal directional pairs: up-down, forward-backward, and sideways. The directions in the sense of to-ward and away-from are directions of movement, which have, as do all directions, a spatial-temporal character.

In locomotion the seeing individual remains in the center of his field of vision, at the hub of the directional spokes proceeding outward from him. But since his Here wanders with him, and space opens up continually before him with new goals, he remains in the middle, unsheltered, in the face of boundless possibilities. With each step toward a goal he loses what now lies behind him. No matter how far he wanders the horizon wanders with him. The tension of the opposition

10. ". . . In this period the world is pure destiny to him, not yet object. Everything has existence for him only insofar as it procures existence to him; what neither gives to him nor takes from him, is not even present to him. . . . Everything that is, is to him through the decree of the moment. . . . Nature lets her rich manifoldness pass before his senses in vain: he sees nothing in her splendid fullness but his booty, in her might and greatness nothing but his enemy. Either he plunges upon the objects and desires to seize them unto himself in his appetite, or the objects press destructively upon him and he thrusts them away from himself in abhorrence and terror. In both cases his relationship to the world of the senses is one of immediate *contact*; and eternally anxious through its pressure, restless and tortured by imperious needs, he nowhere finds rest except in exhaustion and nowhere limits except in exhausted appetite." (Schiller, *Über die asthetische Erziehung des Menschen*, 24. *Brief.*)

does not disappear; the relationship I-world is maintained throughout every change. All aspects are unified in this relationship; the world appears as one and uniform. Directed at the whole and yet simultaneously limited to a place, we experience the other—the world—as the mighty and space as the encompassing. The wide world has been opened to our gaze, but at the same time we have been thrown back on ourselves. We can appropriate only a small part as our own; we can limit and enclose it alone as our possession. For the individual, home and homeland become the center of the world—the place from which he proceeds and to which he returns. In the historical geographic space of human settlement country and city characterize the place, where the individual is at home and belongs.

We cannot grasp the whole of space but we can conceive of it insofar as we are able to cease reflecting upon things only in their relationship to us and their desirability or repugnance to us.[11] Although the observer always remains in the center of his field of vision he can still, as it were, emancipate himself from this position. He can gain distance on things, so that the actuality of seeking for and fleeing from is suspended. The space of action is transformed into a space upon which the observer gazes as though from a tower. "Born to see, bound to behold."[12] He is not immediately touched by the processes in this space. Things lose their demand-character. Then it becomes possible to view them in their relationship to one another before the invariant background of a timeless space—itself abiding, monumental, and homogeneous.

The stage space, on which a play unfolds before our eyes, is merely one variant for realizing the possibility of forming space so that the physical continuity is broken by an ideal boundary. In the theater a role is prescribed not only for the actor, but also for the spectator.[13] He

11. "So long as the human being, in his first physical condition, merely appropriates the world passively, merely senses, then he is still fully one with it, and even, since he himself is merely world, there is for him as yet no world. Only when he, in his aesthetic attitude, places or views the world as outside himself, does his personality sever itself from the world, and does a world appear to him, since he has ceased to constitute one with it." (Schiller, *Uber die asthetische Erziehung des Menschen*, 25. *Brief.*)

12. This line from Goethe's *Faust*, Part II, later served as the title of one of Straus' most beautiful papers: E. Straus, "Born to See, Bound to Behold," *Tijdschrift Voor Filosofie*, 27e Jaargang. 4 [1965] 659–688. (Translator's note.)

13. It is hardly necessary to point out that the words "theater" and "theory," derived from the Greek, stem from the same root, as do the Latin forms taken from them: "spectacle" and "speculation."

must submit to being excluded from any participation in events that occur in his immediate presence. The spectator respects the aesthetic distance and thus allows the stage, which signifies the world, to become the setting for deeds that take place according to their own spatial-temporal lawfulness. The spectator takes his seat at the hour prescribed by calendar and watch. When the stage curtain rises and the stage comes into view, however, he no longer finds himself at the Schiller Place or on Broadway but before the palace of King Oedipus at Thebes, in Wallenstein's camp, or in a rundown warehouse in Mississippi. We follow the poet who conjures up the mythical olden times or a certain day in the "present." We accept that in the dwindling three hours of a theater evening, days and years elapse on the stage. In the 'εποχη (*epoché*) of the spectator we bracket the actuality of our private existence and accept the fictional locality and temporality of the drama. Nevertheless, the eternal coughing of our neighbor disturbs us, and we grab for the program that slides from our lap. Only rarely are we completely removed from our own present through fascination with the play. As spectators we belong to two regions—the fictional and the actual.

Like all forms of fiction aesthetic drama requires a break in the order of time. The actual Now of seeing and the temporal order of what is seen must be differentiated from one another in the encounter of the spectator with the play. In German linguistic usage one speaks of *Vorstellung*, a presentation or placing before. The presentation of a piece (of what?) is announced or canceled, delayed or repeated. The remarkable aspect of what is presented in such presentations is that it is seen with bodily eyes. During scene changes in the middle of a performance we see the stage crew and their coming and going, just as we have seen the actors in their roles during the actual scenes. The physical and physiological conditions remain the same. Many accoutrements of the closed occidental theater such as the platform with curtain, scenery, and footlights are dispensable ornamentation. Their superfluousness is proved by the classical Greek theater and the modern arena theater in which the seats are arranged around a stage open on all sides. In spite of almost immediate proximity with the actors, the audience, in the attitude of spectator, is able to establish successfully even here the aesthetic distance of space and time. The conversations and actions of the play are understood as such. The figures on the stage tell us who and what they are; from their mouths we learn the place and time of the action. The language reveals the sense of the events occurring before the eyes of the spectator. Through its general meaning the word defines the concrete details of the visible proceedings. The particular receives its sense from the general. The spectator resembles someone who is hypnotized, who allows the hypnotist to tell him what is occurring around him and to him.

What is seen and heard is not synchronous with the seeing and hearing. The scene of the action, there close before me, is not syn-topic with my seat here in the audience. Two spatial-temporal orders encounter one another and separate from one another. This break, this clear separation of two spatial-temporal orders is possible because in the artificial spatial organization of the theater the natural possibilities of human seeing are enhanced and clarified into the pure forms of onlooking and viewing.[14] The separation succeeds particularly in seeing and hearing—in those modalities of sensory experience that we name the senses of distance—because in them the I-world-relationship is dominated by distance.

It is no accident that astronomy is the oldest among the sciences known to us. The magnificent drama of the starry skies, of the *orbis pictus*, invites astonished wonderment. Stretched out prominently over earthly undertakings, far removed from all human grasp, the firmament encloses the entire world. The unity and uniformity of space is visible in the firmament; man has discovered in it the single order that governs throughout the whole. Thus arises the horror over eclipses, which on a clear night with a full moon appear to threaten the secure course of the stars. The accidental has a place in the sublunar world; yet exceptions only verify the rule that governs above in rigorous perfection. The rainbow, which stretches out in a display of brilliant colors from heaven to earth in its complete form, becomes a sign of cosmic order in the biblical tale. Understood as a heiroglyphic of human history, the promise to Noah points to a decisive transformation in the conception of nature. God renounces his power for violent interventions in natural events. With God confined to His place, the thinking human being was

14. In the everyday world the aesthetic phenomenon and the physiognomic object, the charming and the enticing, the beholding and the seeing, are so intimately connected, that in the region of human eroticism desire is determined and guided by the beautiful. A girl, attractive in the sense of "charming," is attractive simultaneously in the sense of "sex-appeal." The ambiguity of the word "stimulus" (*Reiz*), which can be understood and misunderstood as excitation, arousal, or stimulus, has contributed much to the confusion in physiology and psychology.

able to seize upon the order of the whole, to represent it in a schema, to reproduce it in a picture. The constellations guided his reflecting. A smell or a noise may indicate something else; a representation may represent something absent. The picture, however, is a likeness of a primal picture; in the visible it imitates the visible; in artistic form, a natural formation. It brings a distant formation into tangible nearness. Although pictures may have served magical purposes, nevertheless magic did not generate them; it only seized upon them. The possibility of the likeness is given as soon as, in the aesthetic-contemplative attitude and distance, the things are themselves viewed and understood in their formal organization (Gestalt). It is this dematerialized Gestalt that can be reproduced in another, essentially indifferent, substance and measure.

Psychology distinguishes surface colors from other modes of appearance in the world of color. Color world and tactile world—these are beautiful but perhaps not very useful expressions, because they mislead one to the notion that the color world and the other sensory worlds are independent constituents of the world, that the world is joined together out of such partial worlds. In the world, however, surface colors do not appear to us; we see colored surfaces. The surface, this facade, hides and veils, but it also indicates and expresses what is concealed under and behind it. It is the surface of a hidden depth that appears through it. Looking at the things removed from our grasp cannot penetrate their surface. In the surface we encounter a mystery that we endeavor to solve. Either we seek to remove the surface as a troublesome, covering hull in order to understand how the object is composed, or we understand the surface as an expression, as a superficies,[15] that indicates what takes place in the interior. The one way leads to scientific analysis; the other, to pictorial representation. Accordingly the picture is no mere imitation, no likeness of a limited view; rather it represents the nature of the object more clearly than the original, just as we think—for example, when we speak of a "picture of terror" or a "picture of health"—that the tangible reality appears with the pregnancy and expressive power of a picture. It is precisely this that also turns the drama into the play; this is why the material orders of the events are suspended in their weightiness. Thus the poet

15. The English word "surface," corresponding to the Latin *superficies*, is fuller than the meagre German expression *Ober-flache*. Face and *-ifices* are derived from *facies*, which itself is probably related to the Greek. (Translator's note: *facies* has the meaning "a face.")

disposes freely but meaningfully with the spatial and temporal relationships, he arranges the fragments of the real happening into a coherence more comprehensible than reality. Belonging not to the moment, the play is always of importance at the present time. In its suprahistorical significance it addresses us and moves us. "The purpose of playing whose end both at the first and now was and is to hold as 'twere the mirror up to Nature. To show virtue her own feature, scorn her own image and the very age and body of the time, his form and pressure" (Hamlet III, 2).

Even the schematic representation of a map, for example, is no imitative abstraction. It does not repeat in simplification that which, in the moment, was accessible all along. Rather, breaking through the horizon of the visible, it unites fragments for the first time into a meaningfully arranged whole. It is a compositing abstraction. Disposing freely with scale, the observer makes possible the representation of extreme largeness and smallness in a size suitable to human visual relationships: the earth as an easily managed globe, atomic structure in a viewable model. Finally, the observer succeeds in defining his own position reflexively, as a part in the whole. Moving within his space of action, continuously bound to perspectival vision, he succeeds in projecting a design in which all those perspectives that dominated during the measuring have been eliminated, and the results of the measurement carried out in the course of time have been introduced as timeless relationships. As the observer now defines his own position by such a plan, he transforms himself from the viewer into the viewed and sees himself as if in a mirror. The physical-optic reflection thus for the first time becomes a mirror-image, in that we knowingly see ourselves as we appear to others. The observer plays a double role, he is simultaneously measurer and measured. Nevertheless he cannot exchange the two roles at any time he wishes. Only from his original position in the world in contraposition to things, only with a view to the whole of space, is man able to conceive space mathematically and to take its measure in all directions. Only in the whole of space is a coordinate system constructable, definable in its layout, and measurable in its intervals. Only from the central position of the observer is the space-whole comprehensible. The side-by-side relations of the spatial order cannot be conceived from within the side-by-side relations.

10 / *Measurements*

Now is the time to ask: can we, and how can we, ascribe the accomplishments of the measuring observer and of his brain to *the* brain?

The brain does not communicate to the observer the spatial relation of contraposition. It does not stretch out beyond itself, and make possible the comprehension of another body in space. Since the brain has no relationship to the totality of space, it also cannot define its own place in space. It does not have a *Here* at its disposal to make the departure point of an action. It cannot travel. In the locomotion of an organism the brain is carried along, not in the manner of a traveler in a vehicle, but rather like a suitcase in a luggage van. If, in order to view the "seven wonders of the world," someone wanders from one country to another, the brain remains in its place on the entire journey, mediating between input and output. It sends impulses to muscle fibres, but it has no relation to the path. All cerebral processes, regardless of kind and order of magnitude, are *in* the brain and are localized there only. Circumscribed to its place in the cranium, *the* brain is affected and excited by stimuli. A stimulus has no independent existence comparable to that of a visible object. Objects are not stimuli. The light reflected by an object becomes a stimulus only when it has reached the retina and has been taken up into the sensing system. The physiologist can make a brain the object of his observation; from his position opposite the brain, he can place it next to another and compare the two. While he objectively views the two objects A and B in their relationship to one another, the brain of the observer has been excited by light rays reflected from A and B. Nevertheless, the relationship between these optic stimuli and the central processes dependent on them by no means corresponds to the relationship, visually comprehended by the observer, of the objects among themselves, or to the relationship of the observer to them. The spatial relationship of one brain to another, or to any physical body for that matter, is that of side-by-side-ness. The brain is "something at hand" next to another "something at hand," but it cannot itself comprehend this relationship.

In the attempt to explain the behavior and the experience of animal and human by brain processes, the physiologist finds himself forced to reinterpret the spatial-temporal relations of the observer with his environment into the topographic and chronological relations of cerebral processes. Simultaneity and succession of nervous excitations become the ultimate explanatory principles. Nevertheless, factual simultaneity is not experienced simultaneity, and this is not equivalent with a known temporal connection. The concept of simultaneity has been of the utmost importance for the foundation of the theory of relativity. Einstein turned his attention to the physical presuppositions of the determination of the simultaneity of events. He recognized that the observer is no uninvolved spectator, that the aesthetic distance cannot suspend the physical continuity. In the determination of simultaneity the observer is himself engaged in his own corporeality. However, the clarification of the physical conditions still always takes for granted the psychological possibility of the comprehension of simultaneity. Simultaneous experiences are separate and different processes, which can be united only with respect to a third. The unifying third is the "Now" of the observer; events are simultaneous insofar as, for the observer, they belong to the same moment of his becoming. Simultaneous events coincide in a personal Now of experiencing. The observer unites them in a single moment of his experience. Whether or not his judgment is justified physically, is psychologically irrelevant. We can observe simultaneity because the world opens before us in many directions or converges on us from many directions. The observer is the temporal and spatial center, within which the unification takes place, but in such a way that the processes, combined in *a* Now nevertheless remain separate, one from the other. Experienced simultaneity does not yet signify congruence because it does not suspend the difference and separateness. Known simultaneity is also the condition for all distinguishing.

In a certain sense it is also the condition for the comprehension of the succession of events, of the earlier and the later. When we determine a relationship of cause and effect both processes—the earlier and the later—appear as members of one temporal relation. We draw them together, yet leave them separated in their places. Should the sequence of processes be recorded photographically, the earlier takes a different place on the reel of film than the later. They can never come together; the one has disappeared when the other emerges. As individual processes they can never comprehend their relation to one another. The order of the world comprehended in experience is not a mirror image of the order of experiences. Neither through a mere succession of impressions nor through the succession of cerebral processes presumably underlying these impressions, could we obtain the impression of succession. The things we experience come to be known as parts of an encompassing whole, as individual but not as isolated.

Like factual simultaneity, factual succession is also not identical with the succession that is experienced, and the experienced sequence is not identical with the known succession. In using a microscope we change from a weak to a stronger enlargement, and then turn back to the weaker. The impressions follow one another; they are experienced one after another. We nevertheless do not presume that the intimate structure of the tissue made visible through the microscope succeeds the coarser relationships; rather we understand both as enlargements of the spatial structure of the section on the slide. The microscope makes the immanent structures visible to us one after another. The microscope is a complicated apparatus and this optical instrument could only have been invented and made effective because the possibility of gaining insight into simultaneous relationships in the succession of impressions is a human heritage. The microscope—a most ingenious realization of this possibility—has not generated the possibility. In the laboratory we make use of the microscope; in the theater we use an opera glass; in everyday life we draw nearer to an object until we have reached the optimal optic distance. Even in such an "attitude"[16] the impressions follow one after another. In their succession they enable us to see the same object in changing degrees of distinctness.

It is advisable to think through this situation further with the simplest possible examples. In removing an egg shell we see the shell, the egg white, and the yolk, one after another. The impressions follow one another. We advance from the shell to the core without doubting for a moment that, in the object we look at, the shell and the core are

16. By quotation marks Straus emphasizes the derivation of the German *Einstellung* ("attitude"), from *stellen*, or "to place." In an attitude we place ourselves in relation to something (translator's note).

simultaneous.[17] We take a letter out of an envelope in order to read the lines and thus to perceive the sense of the message. In the first phase of such an everyday happening we understand something simultaneous in the succession of impressions; in the second phase we let our glance wander over the lines and thereby transform the simultaneity of letters and lines into a succession of words and sentences. The lines of a letter or the leaves of a book no more follow one after the other than do the leaves of a tree or the keys of a piano. We can, however, touch them and move them one after another. Motion and rest, physically understood, are relative concepts. A pointer swings in front of a scale, the scale is at rest; the pointer, in motion. However, we see both the scale and the pointer simultaneously—the at-rest and the moving. In the everyday act of going we move ourselves in relation to an environment at rest. We see the houses and the streets at rest as well as ourselves in motion. The opposition between the succession of impressions and the simultaneity of the objective order does not coincide with that of immanent and transitive or of phenomenal and transcendent time. In sensory experiencing, in the I-world-relationship, two temporal orders encounter one another and are differentiated: that of my own becoming and that of the world. I experience my Now as a moment in the occurrence of the world.

Sensory experience distinguishes man and animal before all other natural forms. It is the experience of moving creatures. Mobility places us in contraposition to everything else. In this contraposition sensory experience is possible for the first time. Accordingly, sensory experience in its nature and content, in its limitation and its possibilities, is defined by this being-mobile. Mobility requires latitude, in which the abode *Here* is exchangeable with other places *There*—as so many possible goals. Accordingly, in sensory experience we experience the space of the world as the encompassing. We experience ourselves in the world as parts in opposition to a whole, as wandering and changeable in opposition to that which remains still. Only in particular encounters can we approach the whole and the all. In sensory experience the whole shows itself in partial views; it shows itself in perspectival views and therefore in perspectival distortion. Sensory experi-

17. In mathematics the proof follows the proposition. In the discursive consideration of the reasons, we make the truth of the theses sensible for ourselves. In our "train of thought" we follow the demonstration of proof step by step, knowing well that the inference is not later than the reason.

ence is at any given time my experience; it is co-constituted and delimited through my personal standpoint. It is indeed my experience, but it is my experience of the world. As mobile beings we can exchange our standpoint with other places. Nevertheless, in the change of particular positions the basic relationship of being-mobile remains unchanged, as does the relationship of the encompassed to the encompassing, of the part to a whole, of the changeable to the abiding. In the change of position, in the succession of temporal impressions, the world shows itself to me as one and the same. The perspectival views are transparent for the gaze directed toward the whole and the enduring.

In my encounters with the world I find myself constantly in the center; correspondingly, a partial view moves into the focus of my gaze. The organization of the visual field—its division into zones of central and peripheral seeing—perhaps illustrates and actualizes at the fullest the I-world-relation of sensory experience. That which appears clearly and distinctly is not isolated. It appears as part, rendered prominent within a horizon of the visible, surrounded by what is not yet but which can possibly become seen clearly. Language is rich in expressions that describe the relationship—the transition from peripheral to central seeing—as an active achievement, for example, to let one's gaze rest on something, to let one's gaze wander, to cast a glance, or an eye, at something. But language is no less rich in images that express the same relationship as something I undergo. We say that things stand out; they fall into, even leap into sight; they capture our glance. We speak of impressions; the receptivity of sensory experience is thereby interpreted as passivity. Now it is certainly true that things assault our glance: as soon as I open my eyes I must see objects exactly as they show themselves to me; I have no choice. Nevertheless, a captured glance is indeed actually bound, but on the other hand it is potentially free. The coercion is experienced as coercion because in the actual surprise we discover the other in its power; but at the same time we also discover it in a delimitation conditioned by our mobile standpoint. In sensory experience the I-world-relationship of a mobile being manifests itself in its continuity. Sensory experience is not joined together out of many individual impressions. The partial views of the world accessible in a delimited moment and from a delimited standpoint are experienced as parts, which refer to a connection with other parts. In each phase of the I-world-relation we comprehend—

already preconceptually—the possibility of the significant advance from one moment to the next. Here and there, now and then, center and periphery stand in a relationship of sense, which cannot be exhaustively represented mathematically as relation of part to part and part to whole.

Measuring means to determine a proportion between the measure chosen as unit and the unknown size of the line that is to be measured. The interval A-B amounts to a kilometer. In linguistic abbreviation and condensation this signifies: The length of the route from A to B is in a ratio of 1,000:1 with a standard meter measure. In the simple practice of measuring, the meter measuring stick as a fixed physical body is compared with the geographic formation of the route. In theory the abstract extension of the one is set in relation to the abstract length of the other.

The act of measuring requires time. On the other hand the length of a kilometer that is arrived at—the relationship 1000:1—is understood as a timeless relationship. In order to measure a distance the measurer must himself proceed from A to B, that is, he must be mobile. He may not be chained to the place A as to a point in geometric space, or like a rock or a tree in physical space. He must be capable of establishing contact with B at a distance, over the separating interval A-B.

Distance requires connection. Let us assume A and B to be the end points of a line that is to be measured. The two places—A marked by a stone; B, by a tree—have nothing to do with one another. The interval of a mile separates them, one from the other; nevertheless the measurer connects them and fixes them—the separated places—as endpoints of a line: a mile altogether. In the measuring the separation is overcome yet nevertheless remains as an interval in the order of side-by-side relationships, which is also the order of external one-to-another relationships.

A thousand paces (*milia passuum*) is an expression that still describes the result of the measuring—the arrived-at length—by an implicit reference to the act of measuring. The word "mile" tells the uninformed nothing of its derivation from *milia*. He reckons with the mile as a measure of length, in which the 1,000 paces have been understood together in the continuity of a single line. In pacing off this route we count the steps, or more precisely the stepping motions. The motions follow one after another; the one obliterates the other; they form no whole; they elapse as we go. Nevertheless, as a traversed line the steps arrange themselves together in a row; taken together they form the route that we traverse. The stepping motions are equivalent to step lengths. The steps, however, each cover a particular interval of the total line. Since we associate the swinging stepping motions with the counting series, we draw them together in their temporal succession into a group of the timeless unity, "thousand." This serves simultaneously to divide the way into parts and to assemble it again out of parts. The unity "thousand" is understood as the plurality $1 + 1 + 1 + 1 \ldots = 1000$. The measuring is thus a process both of combination and of partition.

In the process of measuring, A was the start and B was the goal of the wandering. After the measuring is complete the two separated places are understood as temporally indifferent endpoints of a line. The measuring began at A and ended at B. For the active individual A was earlier than B. Proceeding from the place A he directed himself toward the place B. In the totality of space at rest, the place B lay visible in the present, yet also lay before him as goal, as a future abode. As endpoints of the measured line, A and B are simultaneous. In the act of measuring, two orders of time encounter each other.

The polarity of the I-world-relation makes active adaptation possible, because active adaptation in any form—whether as subjection under existing conditions or as domination over them—always presupposes a relation of correspondence. In experiencing, in the encounter with the world the animal also comprehends two orders. Yet because "everything only has existence for it, insofar as it provides existence to it," its sentient understanding remains at the isolated phase, restricted to the transition from one part to the next; it learns to know the world only "by rote." The human being, however, because "he escapes the decree of the moment," is capable of surpassing that accomplishment, and of comprehending the interconnections of the parts both to one another and within the whole. In experiencing, the world is the encompassing, the mighty, the enduring. However, in his mobility the experiencing individual is given a limited power over the parts. He can combine these parts systematically according to the order lurking in the whole, for his own purposes. The human being becomes the artisan (*homo faber*) on the basis of the polarity of the I-world-encounter.

The legend tells that Archimedes, after the discovery of the laws of leverage, boasted that he desired a place on which to stand, from which he could turn the world upside down. He failed to realize that this wish had already long been fulfilled for the observing, discovering, inventing human being.

Index

Single experience, 50
Sinnentnahme. See Sense derivation
Socrates, ix
Somatic medicine, 90
Soviet Russia, 39
Space
  Euclidean, 149
  visual, 150
Spatiality, ix, 152, 159
Specific concretization, 31
Spectres, 30
Spencer, Herbert, 26
Spiegelberg, Herbert, viii, ix
Spiessburger, 85
Spinozistic pantheism, 143
Squandering, 43
Stimulus, 17
  vs. event, 99
Stimulus-increment, 17, 18, 22
Stimulus-response theory, xiv, 19
Stimulus-threshold, 100
Straus, Caesar, vii
Straus-Negbaur, Antonie, vii
Stream experiencing, 46
Stumpf, Carl, vii, 59
Subjective readiness, 78, 79, 80, 91
Subjective reality, 88
Subjective representation, 79
Succession, 161
  factual, 160
Suchness (*sosein*), 50
Sudden, the, 35, 98, 99
Superego, 90, 129
Sylvester's Eve, 132
Symptoms
  anxiety-neurotic, 62
  organ-neurotic, 114

Taboo-sensibility, 89
Temporality, ix
Tensions, 56, 81
Tertium comparationis, 51, 52
Thaetatus, ix
Time
  experience-immanental, 102, 105
  Newtonian, xv
  transitive, 102

Time-structure of experience, 19
Totality (wholeness), 84
  See also Gestalt
Trace reflexes, 67, 71
Transformation of experience, 33, 34, 36
Transition, 71
Transitive time, 102
Trauma, 13, 21, 22, 34, 139
  See also Shocking experiences
  birth, 52
  psychic. *See* Psychic trauma

Umwelt, 59
Unconscious, 19, 127
Universal concretization, 31
Unrechts. See Injustice
Untouchedness, 77
Upright posture, x, xv
"Upright Posture, The," xvi

Value-actualizations, 120
Value-attitudes, 120
Vasomotor changes, 94
Visual space, 150
Vom Sinn der Sinne, ix
Von Gebsattel, Victor, viii, ix
Von Weizacher, F., 115, 116, 117
Vorstellung, 154

Weber-Fechner law, 57, 58
Weimar Republic, viii
Werner, H., 49
Wertheimer, Leo, viii
Wesenheiten. See Existentiell
Wholeness (totality), 84
  See also Gestalt
Work, 118
World edifice, 25
Wundt, Wilhelm, 58

Yearning, 80

Zutt, Jung, viii, ix